MEMOIRS
of the
American Mathematical Society

ISSN 0065-9266 (print)
ISSN 1947-6221 (online)

Number 1499

Homotopy Fibrations with a Section After Looping

Stephen Theriault

July 2024 • Volume 299 • Number 1499 (sixth of 7 numbers)

AMERICAN MATHEMATICAL SOCIETY

Library of Congress Cataloging-in-Publication Data

Cataloging-in-Publication Data has been applied for by the AMS.
See http://www.loc.gov/publish/cip/.
DOI: https://doi.org/10.1090/memo/1499

Memoirs of the American Mathematical Society

This journal is devoted entirely to research in pure and applied mathematics.

Subscription information. Beginning in 2024, Memoirs will be published monthly through 2026. Memoirs is also accessible from www.ams.org/journals. The 2024 subscription begins with volume 293 and consists of twelve mailings, each containing one or more numbers. Individual subscription prices for 2024 are as follows. For electronic only: US$1013. For paper delivery: US$1,163. Add US$22 for delivery within the United States; US$150 for surface delivery outside the United States. Upon request, subscribers to paper delivery of this journal are also entitled to receive electronic delivery. For information on institutional pricing, please visit `https://www.ams.org/publications/journals/subscriberinfo`. Subscription renewals are subject to late fees. See `www.ams.org/journal-faq` for more journal subscription information. Each number may be ordered separately; please specify number when ordering an individual number.

Back number information. For back issues see `www.ams.org/backvols`.

Subscriptions and orders should be addressed to the American Mathematical Society, P. O. Box 845904, Boston, MA 02284-5904 USA. *All orders must be accompanied by payment.* Other correspondence should be addressed to 201 Charles Street, Providence, RI 02904-2213 USA.

Memoirs of the American Mathematical Society (ISSN 0065-9266 (print); 1947-6221 (online)) is published bimonthly (each volume consisting usually of more than one number) by the American Mathematical Society at 201 Charles Street, Providence, RI 02904-2213 USA. Periodicals postage paid at Providence, RI. Postmaster: Send address changes to Memoirs, American Mathematical Society, 201 Charles Street, Providence, RI 02904-2213 USA.

This publication is indexed in *Mathematical Reviews®, Zentralblatt MATH, Science Citation Index®, Science Citation Index™-Expanded, ISI Alerting Services℠, SciSearch®, Research Alert®, CompuMath Citation Index®, Current Contents®/Physical, Chemical & Earth Sciences.*
This publication is archived in *Portico* and *CLOCKSS.*
Printed in the United States of America.

∞ The paper used in this book is acid-free and falls within the guidelines established to ensure permanence and durability.
Visit the AMS home page at `https://www.ams.org/`

10 9 8 7 6 5 4 3 2 1 29 28 27 26 25 24

Contents

Abstract

We analyze a general family of fibrations which, after looping, have sections. Methods are developed to determine the homotopy type of the fibre and the homotopy classes of the map from the fibre to the base. The methods are driven by applications to two-cones, Poincaré Duality complexes, the connected sum operation, and polyhedral products.

Received by the editor May 23, 2020, and, in revised form, August 11, 2021, and February 8, 2022.

Article electronically published on July 11, 2024.

DOI: https://doi.org/10.1090/memo/1499

2020 *Mathematics Subject Classification.* Primary 55P35, 57N65; Secondary 55Q15.

Key words and phrases. Fibration, cofibration, two-cone, Poincaré Duality complex, connected sum, polyhedral product.

CHAPTER 1

Introduction

A fundamental goal in homotopy theory is to determine the homotopy types of spaces and the homotopy classes of the maps between them. This paper builds on new methods developed in [**BT2**] in order to do that in an appropriate context. The applications are wide-ranging, informing on the homotopy theory of two-cones, Poincaré Duality complexes, connected sums, and polyhedral products.

To describe the context, it will be assumed throughout that all spaces are CW-complexes so that weak homotopy equivalences are homotopy equivalences. Suppose that there is a homotopy fibration $E \xrightarrow{p} Y \xrightarrow{h} Z$ and a homotopy cofibration $\Sigma A \xrightarrow{f} Y \longrightarrow Y'$. Suppose that h extends to a map $h'\colon Y' \longrightarrow Z$ and let E' be the homotopy fibre of h'. This data is assembled into a diagram

$$(1.1)\qquad \begin{array}{ccccc} & & E & \longrightarrow & E' \\ & & \big\downarrow{\scriptstyle p} & & \big\downarrow{\scriptstyle p'} \\ \Sigma A & \xrightarrow{f} & Y & \longrightarrow & Y' \\ & & \big\downarrow{\scriptstyle h} & & \big\downarrow{\scriptstyle h'} \\ & & Z & = & Z \end{array}$$

where the vertical columns and the maps between them form a homotopy fibration diagram. Using either Dold and Lashof [**DL**] or Mather's Cube Lemma [**M**], there is a homotopy pushout

$$(1.2)\qquad \begin{array}{ccc} \Omega Z\times\Sigma A & \longrightarrow & E \\ \big\downarrow{\scriptstyle \pi_1} & & \big\downarrow \\ \Omega Z & \longrightarrow & E' \end{array}$$

where π_1 is the projection. Under favourable circumstances, this homotopy pushout may allow for the homotopy type of E' to be determined, and possibly also the homotopy class of the map $E \longrightarrow E'$. However, much depends on the homotopy class of the map $\Omega Z\times\Sigma A \longrightarrow E$, and this can be difficult to identify with sufficient precision.

Suppose in addition that the map $\Omega Y \xrightarrow{\Omega h} \Omega Z$ has a right homotopy inverse $s\colon \Omega Z \longrightarrow \Omega Y$. Then the homotopy pushout (1.2) simplifies to a homotopy cofibration

$$(1.3)\qquad \Omega Z\ltimes\Sigma A \xrightarrow{\theta} E \longrightarrow E'$$

for some map θ. In the special case when $Y' = Z$ and h' is the identity map, this implies that E' is contractible so θ is a homotopy equivalence. But in general this cofibration by itself says little about the precision with which θ can be identified.

However, as will be explained in Chapter 2, the existence of a right homotopy inverse for Ωh implies that there is a profound connection between θ, the homotopy action of ΩZ on E, and Whitehead products mapping into Y. Specifically, there is a homotopy commutative diagram

$$\begin{array}{ccc} \Omega Z \ltimes \Sigma A & \xrightarrow{\theta} & E \\ \big\downarrow \simeq & & \big\downarrow p \\ (\Omega Z \wedge \Sigma A) \vee \Sigma A & \xrightarrow{[\gamma, f]+f} & Y \end{array} \tag{1.4}$$

where γ is the composite $\gamma\colon \Sigma\Omega Z \xrightarrow{\Sigma s} \Sigma\Omega Y \xrightarrow{ev} Y$, the map ev is the canonical evaluation map, and $[\gamma, f]$ is the Whitehead product of γ and f. That is, the homotopy class of θ is identified, at least up to composition with p, and this gives a measure of control over the homotopy cofibration $\Omega Z \ltimes \Sigma A \xrightarrow{\theta} E \longrightarrow E'$. But the level of control is often not fine enough to precisely describe the homotopy type of E' in cases of interest. Obtaining that control is the thrust of this paper.

We consider, then, homotopy fibrations $E \xrightarrow{p} Y \xrightarrow{h} Z$ which have a section after looping. That is, those for which Ωh has a right homotopy inverse. This begins with a simple but foundational case that will play an important role at many points later on. We move on to consider different families of examples, each of which involves distinctive features that influence how control over θ is obtained.

A foundational case. Consider the homotopy fibration $E \xrightarrow{p} \Sigma X \vee \Sigma Y \xrightarrow{q_1} \Sigma X$ where q_1 is the pinch map to the first wedge summand. Note that q_1 has a right homotopy inverse, so Ωq_1 does as well, implying that this is an example of a homotopy fibration with a section after looping.

For $k \geq 1$, let $X^{\wedge k}$ be the k-fold smash product of X with itself. By [**N3**, Theorem 4.3.2] there is a homotopy equivalence

$$E \simeq \bigvee_{k=0}^{\infty} X^{\wedge k} \wedge \Sigma Y$$

where, by convention, $X^{\wedge 0} \wedge \Sigma Y$ refers to ΣY. Further, let $i_1\colon \Sigma X \longrightarrow \Sigma X \vee \Sigma Y$ and $i_2\colon \Sigma Y \longrightarrow \Sigma X \vee \Sigma Y$ be the inclusions of the first and second wedge summands respectively. Let $ad^0(i_1)(i_2) = i_2$ and for $k \geq 1$ let $ad^k(i_1)(i_2)$ be the Whitehead product $[i_1, ad^{k-1}(i_1)(i_2)]$. Then [**N3**, Theorem 4.3.2] shows that, under the homotopy equivalence for E above, the map p may be identified as $\bigvee_{k=0}^{\infty} ad^k(i_1)(i_2)$.

We give an alternative proof of this which has the advantage of having a compatibility with the map θ in (1.4). Here, the general homotopy cofibration $\Sigma A \xrightarrow{f} Y \longrightarrow Y'$ specifies to $\Sigma Y \xrightarrow{i_2} \Sigma X \vee \Sigma Y \xrightarrow{q_1} \Sigma X$ and θ takes the form of a map $\Omega\Sigma X \ltimes \Sigma Y \xrightarrow{\theta} E$. The point to emphasize is that our choice of a homotopy equivalence for E has the additional property of respecting the homotopy action of $\Omega\Sigma X$ on E.

THEOREM 1.1 (appearing in the text as Theorem 3.15). *Let X and Y be path-connected, pointed spaces and consider the homotopy fibration $E \longrightarrow \Sigma X \vee \Sigma Y \xrightarrow{q_1}$*

ΣX. *There is a homotopy commutative diagram*

$$\begin{array}{ccc} \bigvee_{k=0}^{\infty} X^{\wedge k} \wedge \Sigma Y & \xrightarrow{\ \mathfrak{d}\ } & E \\ & \searrow^{\bigvee_{k=0}^{\infty} ad^k(i_1)(i_2)} & \big\downarrow p \\ & & \Sigma X \vee \Sigma Y \end{array}$$

where:

(a) $\mathfrak{d}$ *is a homotopy equivalence;*
(b) $\Sigma\mathfrak{d} \simeq \Sigma d$, *where* d *is the composite*

$$\bigvee_{k=0}^{\infty} X^{\wedge k} \wedge \Sigma Y \xrightarrow{c} \Omega\Sigma X \ltimes \Sigma Y \xrightarrow{\theta} E.$$

Note that the maps $\mathfrak{d}$ and d may not be homotopy equivalent but their suspensions are. Consequently, they induce the same map in homology. As $\mathfrak{d}$ is a homotopy equivalence, it induces an isomorphism in homology, and therefore so does d, and hence d is also a homotopy equivalence by Whitehead's Theorem. This ability to use one homotopy equivalence to prove the existence of another will be used repeatedly throughout. It has the advantage of allowing us to exchange maps that have different properties: in this case the maps $\mathfrak{d}$ behaves well with respect to Whitehead products while the map d (via c) behaves better with respect to the multiplication on $\Omega\Sigma X$.

Two-cones. A two-cone is the homotopy cofibre C of a map $\Sigma A \longrightarrow \Sigma B$ where A and B are both path-connected. More generally, one could consider a map between co-H-spaces instead of suspensions, but the latter simplifies the exposition. This notion can be iterated: a finite CW-complex X has *cone-length* t if t is the smallest number such that there is a sequence of homotopy cofibrations $\Sigma A_k \longrightarrow C_{k-1} \longrightarrow C_k$ for $1 \leq k \leq t$ where C_0 is some initial space ΣA_0 and $C_t \simeq X$. Cone-length is an upper bound on the Lusternik-Schnirelmann category of X. A great deal of work has gone into studying cone-length (see [**CLOT**] for a comprehensive overview). The homotopy theory around two-cones and their based loop spaces has received particular attention [**A, FHT, FT2**] since they are the nearest neighbour to suspensions, whose based loop spaces are well understood through the Bott-Samelson Theorem, the James construction, and the Hilton-Milnor Theorem.

In Theorem 4.6 we prove a general result which lets us consider, as examples, certain families of two-cones. One case is the following. Define the two-cone M_k by the homotopy cofibration

$$\Sigma X^{\wedge k} \wedge \Sigma Y \xrightarrow{ad^k(i_1)(i_2)} \Sigma X \vee \Sigma Y \longrightarrow M_k.$$

We give a homotopy decomposition of ΩM_k. Note that as $ad^k(i_1)(i_2)$ is an iterated Whitehead product, it composes trivially with the pinch map $\Sigma X \vee \Sigma Y \xrightarrow{q_1} \Sigma X$, implying that q_1 extends to a map $M_k \xrightarrow{q'} \Sigma X$. Define the map γ_k by the composite

$$\gamma_k \colon \bigvee_{t=0}^{k-1} X^{\wedge t} \wedge \Sigma Y \xrightarrow{\bigvee_{t=0}^{k-1} ad^t(i_1)(i_2)} \Sigma X \vee \Sigma Y \longrightarrow M_k.$$

THEOREM 1.2 (appearing in the text as Theorem 4.9). *For $k \geq 1$, there is a homotopy fibration*

$$\bigvee_{t=0}^{k-1} X^{\wedge t} \wedge \Sigma Y \xrightarrow{\gamma_k} M_k \xrightarrow{q'} \Sigma X$$

which splits after looping to give a homotopy equivalence

$$\Omega M_k \simeq \Omega\Sigma X \times \Omega(\bigvee_{t=0}^{k-1} X^{\wedge k} \wedge \Sigma Y).$$

Particular examples of interest occur when X and Y are both spheres or Moore spaces. These are discussed in Chapter 4; they give a large family of examples that satisfy Moore's conjecture.

Poincaré Duality complexes. A finite CW-complex X is a *Poincaré Duality complex* if $H^*(X;\mathbb{Z})$ satisfies Poincaré Duality. These spaces are generalizations of closed, orientable manifolds. Poincaré Duality complexes have a long history in both geometry and topology (see the survey by Klein [**K**]) and recently there has been progress in analyzing their homotopy groups through homotopy decompositions of their loop spaces. In particular, Beben and Wu [**BW**] studied $(n-1)$-connected $(2n+1)$-dimensional Poincaré Duality complexes M with n odd, $n \geq 6$ and $H_{2n-1}(M;\mathbb{Z})$ consisting only of odd torsion; Beben and the author [**BT1**] studied all $(n-1)$-connected $2n$-dimensional Poincaré Duality complexes; this case was also considered using different methods by Sa. Basu and So. Basu [**BB**], and Sa. Basu [**Ba**] went on to consider $(n-1)$-connected $(2n+1)$-dimensional Poincaré Duality complexes M with $H_n(M;\mathbb{Z})$ having at least one integral summand. In [**BT2**], Beben and the author developed the new methods that are the basis of this paper and used them to recover in a unified way the results in [**Ba, BB, BT1**].

The case of an $(n-1)$-connected $(2n+1)$-dimensional Poincaré Duality complex M when n is even and $H_n(M;\mathbb{Z})$ consists only of odd torsion is trickier. The methods used in [**BW**] do not work. They showed that if n is odd then there is a space V and a map $M \xrightarrow{h} V$ where Ωh has a right homotopy inverse and, for an appropriate prime p, $H_*(V;\mathbb{Z}/p\mathbb{Z}) \cong \Lambda(x,y)$ with $|x| = n$, $|y| = n+1$ and x and y connected by a Bockstein (possibly of higher order). No such space exists when n is even. The problem boils down to the following. For a prime p and integers $r \geq 1$ and $m \geq 2$, the *mod-p^r Moore space* $P^m(p^r)$ is the homotopy cofibre of the degree p^r map on S^{m-1}. It is characterized by the fact that $\widetilde{H}_n(P^m(p^r);\mathbb{Z})$ is $\mathbb{Z}/p^r\mathbb{Z}$ if $n = m$ and is 0 if $n \neq m$. The factor of least connectivity in $\Omega P^{2n}(p^r)$ is the homotopy fibre of the degree p^r map on S^{2n-1}, which does retract off ΩV for a certain 3-cell complex V, but the factor of least connectivity in $\Omega P^{2n+1}(p^r)$ is a space constructed by Cohen, Moore and Neisendorfer [**CMN**] whose mod-p homology is much more complex and is not a factor of ΩV for some 3-cell complex V. So we approach the problem from a different perspective. Instead of trying to find a factor of least connectivity that is indecomposable, we are content to find a copy of $\Omega P^{n+1}(p^r)$ in ΩM and aim to identify the complementary factor.

In doing this we consider a much larger family of examples, most of which are not Poincaré Duality complexes. A general result is proved in Theorem 5.8 which is then increasingly specialized. In the case presented below, the attaching map f

in a homotopy cofibration $S^{2n} \xrightarrow{f} \bigvee_{i=1}^{m} P^{n+1}(p^r) \longrightarrow M$ factors through Whitehead products and so composes trivially with the pinch map $\bigvee_{i=1}^{m} P^{n+1}(p^r) \xrightarrow{q_1} P^{n+1}(p^r)$ to the first wedge summand. Therefore q_1 extends to a map $M \xrightarrow{q'} P^{n+1}(p^r)$. For $1 \leq k \leq m$, let

$$i_k \colon P^{n+1}(p^r) \longrightarrow \bigvee_{i=1}^{m} P^{n+1}(p^r)$$

be the inclusion of the kth-wedge summand. Note that the Whitehead product $[i_j, i_k]$ is a map $\Sigma P^n(p^r) \wedge P^n(p^r) \longrightarrow \bigvee_{i=1}^{m} P^{n+1}(p^r)$. There is a map $S^{2n} \xrightarrow{v} \Sigma P^n(p^r) \wedge P^n(p^r)$ which induces an injection in mod-p homology.

THEOREM 1.3 (appearing in the text as Theorem 7.7). *Let p be an odd prime, $r \geq 1$ and $n \geq 2$. Suppose that there is a homotopy cofibration*

$$S^{2n} \xrightarrow{f} \bigvee_{i=1}^{m} P^{n+1}(p^r) \longrightarrow M$$

where $f = \sum_{1 \leq j < k \leq m} [i_j, i_k] \circ (d_{j,k} \cdot v)$ for $d_{j,k} \in \mathbb{Z}$ and at least one $d_{j,k}$ reduces to a unit mod-p. Rearranging the wedge summands $\bigvee_{i=1}^{m} P^{n+1}(p^r)$ so that some $d_{1,t}$ reduces to a unit mod-p, there is a homotopy fibration

$$(\Omega P^{n+1}(p^r) \ltimes \overline{C}) \vee (\bigvee_{i=2}^{m} P^{n+1}(p^r)) \longrightarrow M \xrightarrow{q'} P^{n+1}(p^r)$$

where $\overline{C} \simeq \Big(P^n(p^r) \wedge (\bigvee_{\substack{i=2 \\ i \neq t}}^{m} P^{n+1}(p^r))\Big) \vee \Big(S^{2n+1} \vee P^{2n}(p^r)\Big)$, *and this homotopy fibration splits after looping to give a homotopy equivalence*

$$\Omega M \simeq \Omega P^{n+1}(p^r) \times \Omega\Big((\Omega P^{n+1}(p^r) \ltimes \overline{C}) \vee (\bigvee_{i=2}^{m} P^{n+1}(p^r))\Big).$$

The interpretation of Theorem 1.3 requires care. Some of the spaces M are Poincaré Duality complexes while others are not, and not all $(n-1)$-connected $(2n+1)$-dimensional Poincaré Duality complexes with $H_n(M;\mathbb{Z})$ consisting only of odd torsion have the form described in the theorem. But there are tangible results. For example, simply-connected 5-dimensional Poincaré Duality complexes have been classified by Stöcker [**St**]. The classification shows that if M is a Spin manifold and $H_2(M;\mathbb{Z})$ is a direct sum of $\mathbb{Z}/p^r\mathbb{Z}$'s for p odd, then the attaching map for its top cell has the form described in Theorem 1.3. If M is either a non-Spin manifold or a Poincaré Duality complex that is not a manifold, then the attaching map for the top cell involves a stable term and the theorem does not apply.

Connected sums. A classical problem in homotopy theory is to determine the effect on a CW-complex X, or its loop space ΩX, by attaching a cell. Rational homotopy theory has had some success in this direction for certain families of attaching maps. Let $S^{n-1} \xrightarrow{f} X \xrightarrow{i} X \cup e^n$ be a cofibration where f attaches an n-cell to X and i is the inclusion. The map f is *inert* if Ωi induces an epimorphism in rational homology. This implies that, rationally, Ωi has a right homotopy inverse. Inert maps have received notable attention, for example, in [**FT1, HaL**], as have assorted variants such as nice, lazy and semi-inert attaching maps [**Bu, HeL**].

We consider an integral version of an inert map, and generalize from attaching a cell to attaching a cone, that is, to a cofibration $A \xrightarrow{f} X \xrightarrow{i} X \cup CA$. Modifiying, we consider a homotopy cofibration $\Sigma A \xrightarrow{f} X \xrightarrow{i} X'$ where all spaces are assumed to be simply-connected and have the homotopy type of CW-complexes. The map f is *inert* if Ωi has a right homotopy inverse. Note there is no localization hypothesis here.

Let $\Sigma A \xrightarrow{g} Y \longrightarrow Y'$ be another such cofibration, where g need not be inert. As ΣA is a suspension we may add to obtain $\Sigma A \xrightarrow{f+g} X \vee Y$. In Theorem 8.6 we show that $f+g$ is inert. If C is the homotopy cofibre of $f+g$ then we give a homotopy decomposition for ΩC in terms of X and Y' and prove additional related statements. The property that $f+g$ is inert, regardless of whether g is inert, is intriguing.

As a special case we consider the connected sum $M\#N$ of two Poincaré Duality spaces M and N of the same dimension. It is natural to ask how the homotopy type of $M\#N$ reflects the homotopy types of M and N. Theorem 1.4 provides an answer.

Let X and Y be the $(n-1)$-skeletons of M and N respectively. Then there are homotopy cofibrations $S^{n-1} \xrightarrow{f} X \xrightarrow{h} M$ and $S^{n-1} \xrightarrow{g} Y \xrightarrow{k} N$ that attach the top cells to M and N respectively. The connected sum of M and N has $(n-1)$-skeleton $X \vee Y$ and the attaching map for its top cell is $f+g$. We prove, among other properties, the following.

THEOREM 1.4 (appearing in the text subsumed within Theorem 9.1). *Let M and N be simply-connected Poincaré Duality complexes of dimension n, where $n \geq 2$. Let X and Y be the $(n-1)$-skeletons of M and N respectively. If the inclusion $X \xrightarrow{h} M$ has the property that Ωh has a right homotopy inverse, then the following hold:*

(a) *there is a homotopy equivalence $\Omega(M\#N) \simeq \Omega M \times \Omega(\Omega M \ltimes Y)$;*
(b) *the map $X \vee Y \longrightarrow M\#N$ has a right homotopy inverse after looping.*

In particular, if $X \xrightarrow{h} M$ has the property that Ωh has a right homotopy inverse, then so does $X \vee Y \longrightarrow M\#N$, regardless of whether $Y \xrightarrow{k} N$ has that property. The homotopy fibrations $F \longrightarrow X \xrightarrow{h} M$ and $G \longrightarrow X \vee Y \longrightarrow M\#N$ then both have sections after looping and fit into the framework of the paper.

Interesting examples include connected sums of products of two spheres, which play an important role in toric topology [**BM, GPTW, GIPS**]. Another example would take a connected sum of products of two spheres and take its connected sum with a complex projective space of the same dimension. Many more examples are considered in Chapter 9.

Polyhedral products. Let K be a simplicial complex on m vertices. For $1 \leq i \leq m$, let (X_i, A_i) be a pair of pointed CW-complexes, where A_i is a pointed subspace of X_i. Let $(\underline{X}, \underline{A}) = \{(X_i, A_i)\}_{i=1}^m$ be an m-tuple of CW-pairs. For each simplex (face) $\sigma \in K$, let $(\underline{X}, \underline{A})^\sigma$ be the subspace of $\prod_{i=1}^m X_i$ defined by

$$(\underline{X}, \underline{A})^\sigma = \prod_{i=1}^m Y_i \qquad \text{where} \qquad Y_i = \begin{cases} X_i & \text{if } i \in \sigma \\ A_i & \text{if } i \notin \sigma. \end{cases}$$

The *polyhedral product* determined by $(\underline{X}, \underline{A})$ and K is

$$(\underline{X}, \underline{A})^K = \bigcup_{\sigma \in K} (\underline{X}, \underline{A})^\sigma \subseteq \prod_{i=1}^{m} X_i.$$

For example, suppose each A_i is a point. If K is a disjoint union of m points then $(\underline{X}, \underline{*})^K$ is the wedge $X_1 \vee \cdots \vee X_m$, and if K is the standard $(m-1)$-simplex then $(\underline{X}, \underline{*})^K$ is the product $X_1 \times \cdots \times X_m$.

Polyhedral products are currently a subject of intense study. They are at the locus of several constructions from disparate areas of mathematics: moment-angle complexes in toric topology, complements of complex coordinate subspace arrangements in combinatorics, monomial rings with the Golod property in commutative algebra, intersections of quadrics in complex geometry, and Bestvina-Brady groups in geometric group theory.

An important problem is to study the connection between Whitehead products (and higher Whitehead products) and polyhedral products. There has been significant headway on this in the context of the homotopy fibration

$$(\underline{C\Omega\Sigma X}, \underline{\Omega\Sigma X})^K \longrightarrow (\underline{\Sigma X}, \underline{*})^K \longrightarrow \prod_{i=1}^{m} \Sigma X_i.$$

Here, $(\underline{C\Omega\Sigma X}, \underline{\Omega\Sigma X})^K$ is the polyhedral product that is formed from the pairs $(C\Omega\Sigma X_i, \Omega\Sigma X_i)$ where $C\Omega\Sigma X_i$ is the reduced cone on $\Omega\Sigma X_i$. In a sequence of papers [**GT1, GT2, IK1, IK2**] leading up to K satisfying the combinatorial condition of being totally fillable (this includes shifted complexes and Alexander duals of shellable complexes) the space $(\underline{C\Omega\Sigma X}, \underline{\Omega\Sigma X})^K$ is shown to be homotopy equivalent to a wedge of spaces of the form $\Sigma^t X_{i_1} \wedge \cdots \wedge X_{i_k}$ for various $t \geq 1$ and $1 \leq i_1 < \cdots < i_k \leq m$. In [**GT3, GPTW**] for special cases and [**AP, IK3**] more generally, under such a decomposition the map $(\underline{C\Omega\Sigma X}, \underline{\Omega\Sigma X})^K \longrightarrow (\underline{\Sigma X}, \underline{*})^K$ is a wedge sum of iterated Whitehead products of the form $[v_{i_k}, [\cdots [v_{i_1}, w] \ldots]$ where each v_{i_k} represents the inclusion of ΣX_{i_k} into $(\underline{\Sigma X}, \underline{*})^K$ induced by the inclusion of the vertex i_k into K, and w is a higher Whitehead product corresponding to a (minimal) missing face of K.

We go a step further by showing that Whitehead and higher Whitehead products are pervasive in the formation of the polyhedral products, regardless of whether K is totally fillable. If a set of (minimal) missing faces is attached to K to form a new simplicial complex $\overline{K}$, then we show that on the level of polyhedral products there is a corresponding homotopy cofibration

$$\Sigma A \xrightarrow{f} (\underline{\Sigma X}, \underline{*})^K \longrightarrow (\underline{\Sigma X}, \underline{*})^{\overline{K}}.$$

The inclusion $(\underline{\Sigma X}, \underline{*})^K \longrightarrow \prod_{i=1}^{m} \Sigma X_i$ has a right homotopy inverse after looping, and so fits into the overall framework of the paper. Consider the homotopy cofibration $(\prod_{i=1}^{m} \Omega\Sigma X_i) \ltimes \Sigma A \xrightarrow{\theta} E \longrightarrow E'$ from (1.3). On the one hand, in this context $E = (\underline{C\Omega\Sigma X}, \underline{\Omega\Sigma X})^K$ and $E' = (\underline{C\Omega\Sigma X}, \underline{\Omega\Sigma X})^{\overline{K}}$, and on the other hand, the James construction implies that there is a homotopy equivalence

$$(\prod_{i=1}^{m} \Omega\Sigma X_i) \ltimes \Sigma A \simeq \bigvee_{k=0}^{\infty} \bigvee_{1 \leq i_1 \leq \cdots \leq i_k \leq m} (X_{i_1} \wedge \cdots \wedge X_{i_k}) \wedge \Sigma A.$$

This homotopy equivalence can be chosen so the following holds, where the homotopy between suspended maps reflects the same feature in Theorem 1.1.

THEOREM 1.5 (appearing in the text as Theorem 12.7). *There is a homotopy cofibration*

$$\bigvee_{k=0}^{\infty} \bigvee_{1\leq i_1\leq\cdots\leq i_k\leq m} (X_{i_1}\wedge\cdots\wedge X_{i_k})\wedge\Sigma A \xrightarrow{\zeta} (\underline{C\Omega\Sigma X},\underline{\Omega\Sigma X})^K \longrightarrow (\underline{C\Omega\Sigma X},\underline{\Omega\Sigma X})^{\overline{K}}$$

where the map ζ *has the property that* $\Sigma\zeta\simeq\Sigma\zeta'$ *for a map* ζ' *satisfying a homotopy commutative diagram*

$$\begin{array}{ccc}
\bigvee_{k=0}^{\infty} \bigvee_{1\leq i_1\leq\cdots\leq i_k\leq m} (X_{i_1}\wedge\cdots\wedge X_{i_k})\wedge\Sigma A & \xrightarrow{\zeta'} & (\underline{C\Omega\Sigma X},\underline{\Omega\Sigma X})^K \\
 & \searrow^{\bigvee_{k=1}^{\infty}\bigvee_{1\leq i_1\leq\cdots\leq i_k\leq m}[v_{i_1},[v_{i_2},[\cdots[v_{i_k},f]]\cdots]]} & \downarrow{\scriptstyle p} \\
 & & (\underline{\Sigma X},\underline{*})^K.
\end{array}$$

This paper is organized as follows. Chapter 2 reviews the results in [**BT2**] that will be needed later. Theorem 1.1 is proved in Chapter 3. Chapter 4 then considers two-cones, proves Theorem 1.2, and relates the results to Moore's Conjecture. Chapter 5 proves a general decomposition result in Theorem 5.8 and in Chapter 6 this is specialized and applied to certain families of two-cones. Chapter 7 is a modification of the results in Chapter 6 that leads to the proof of Theorem 1.3 and applications to loop space decompositions of $(n-1)$-connected $(2n+1)$-dimensional Poincaré Duality complexes which are rationally copies of S^{2n+1}. Chapter 8 builds on the notion of an inert map and proves a general decomposition result in Theorem 8.6, while Chapter 9 specializes this to prove Theorem 1.4 and give an array of examples. Chapter 10 turns momentarily to algebra to calculate $H_*(\Omega Y)$ as a Hopf algebra, where

$$E \xrightarrow{p} Y \xrightarrow{h} Z$$

is a homotopy fibration with a section after looping. In Chapter 11 we return to homotopy theory to address a second foundational case involving extensions across the inclusion of a wedge into a product, the James construction and Whitehead products that leads to the explicit description of Whitehead products in toric topology that is stated in Theorem 1.5 and proved in Chapter 12.

It is also useful to have a guide on the sections needed to prove each of the main theorems. Theorem 1.1 appears in Chapter 3 and depends only on Chapter 2. Theorem 1.2 appears in Chapter 4 and depends on Chapters 2 and 3. Theorem 1.3 appears in Chapter 7 and depends on Chapters 2 through 6. Theorem 1.4 appears in Chapter 9 and depends on Chapters 2, 3 and 8. The homological interlude in Chapter 10 depends on Chapters 2 and 3. Theorem 1.5 appears in Chapter 12 and depends only on Chapters 2, 3 and 11.

The author would like to thank the referee for a careful reading of the paper, many valuable suggestions for improvement, and spotting a gap in the original submission, the filling of which led to more interesting mathematics.

CHAPTER 2

Background

This section discusses the ingredients behind Theorem 2.2 as they relate to material that will come later in the paper. Recall the blanket assumption that all spaces are assumed to be CW-complexes so that weak homotopy equivalences are homotopy equivalences. We start with a well known lemma and some notation.

The *left half-smash* of two path-connected spaces A and B is the quotient space defined by the cofibration $A \xrightarrow{i_1} A \times B \longrightarrow A \ltimes B$ where i_1 is the inclusion of the first factor.

LEMMA 2.1. *Let A and B be pointed, path-connected spaces. Then there is a homotopy equivalence*

$$A \ltimes \Sigma B \simeq (A \wedge \Sigma B) \vee \Sigma B$$

which is natural for maps $A \longrightarrow A'$ and $B \longrightarrow B'$. □

For path-connected spaces A and B, let

$$j\colon B \longrightarrow A \ltimes B$$

be the inclusion and let

$$q\colon A \ltimes B \longrightarrow A \wedge B$$

be the quotient map that collapses B to a point. By Lemma 2.1 there is a natural map

$$i\colon A \wedge \Sigma B \longrightarrow A \ltimes \Sigma B$$

which is a right homotopy inverse for Σq.

Suppose that $f\colon A \longrightarrow Y$ and $g\colon B \longrightarrow Y$ are maps. The map from $A \vee B$ to Y determined by f and g is denoted by

$$f \perp g\colon A \vee B \longrightarrow Y.$$

In particular, a choice of the homotopy equivalence in Lemma 2.1 is given by $i \perp j$.

Given maps $a\colon \Sigma A \longrightarrow Y$ and $b\colon \Sigma B \longrightarrow Y$ let $[a, b]\colon \Sigma A \wedge B \longrightarrow Y$ be the Whitehead product of a and b. In what follows we will consider a map $\Omega Z \ltimes \Sigma A \longrightarrow E$ with the property that the composite $\Omega Z \wedge \Sigma A \xrightarrow{i} \Omega Z \ltimes \Sigma A \longrightarrow E$ is a Whitehead product $[\gamma, f]$ for some maps γ and f. As such, in these cases we prefer to write the Whitehead product with domain $\Omega Z \wedge \Sigma A$ rather than $\Sigma\Omega Z \wedge A$ to emphasize the link to the half-smash.

The following was proved in [**BT2**]. Let $ev\colon \Sigma\Omega Y \xrightarrow{ev} Y$ be the canonical evaluation map.

THEOREM 2.2. *Suppose that there is a homotopy fibration $E \xrightarrow{p} Y \xrightarrow{h} Z$ and a homotopy cofibration $\Sigma A \xrightarrow{f} Y \longrightarrow Y'$. Suppose that h extends to a map*

$h'\colon Y' \longrightarrow Z$ and let E' be the homotopy fibre of h'. This data is assembled into a diagram

$$
(2.1)\qquad
\begin{array}{ccccc}
 & & E & \longrightarrow & E' \\
 & & \Big\downarrow{\scriptstyle p} & & \Big\downarrow{\scriptstyle p'} \\
\Sigma A & \xrightarrow{f} & Y & \longrightarrow & Y' \\
 & & \Big\downarrow{\scriptstyle h} & & \Big\downarrow{\scriptstyle h'} \\
 & & Z & = & Z.
\end{array}
$$

where the vertical columns and the maps between them form a homotopy fibration diagram. Suppose in addition that the map $\Omega Y \xrightarrow{\Omega h} \Omega Z$ has a right homotopy inverse $s\colon \Omega Z \longrightarrow \Omega Y$. Then there is a map $\theta\colon \Omega Z \ltimes \Sigma A \longrightarrow E$ such that:

(a) *there is a homotopy cofibration*

$$\Omega Z \ltimes \Sigma A \xrightarrow{\theta} E \longrightarrow E';$$

(b) *there is a homotopy commutative diagram*

$$
\begin{array}{ccc}
\Omega Z \ltimes \Sigma A & \xrightarrow{\quad\theta\quad} & E \\
\Big\downarrow{\scriptstyle (i\perp j)^{-1}} & & \Big\downarrow{\scriptstyle p} \\
(\Omega Z \wedge \Sigma A) \vee \Sigma A & \xrightarrow{[\gamma,f]\perp f} & Y
\end{array}
$$

where γ is the composite $\gamma\colon \Sigma\Omega Z \xrightarrow{\Sigma s} \Sigma\Omega Y \xrightarrow{ev} Y$. □

For the benefit of the reader, and to make explicit parts of the construction that will be used later on, a sketch of the proof of Theorem 2.2 will be given. The key is a link between two seemingly distinct constructions. First, let $\Omega Z \xrightarrow{\partial} E$ be the connecting map for the homotopy fibration $E \xrightarrow{p} Y \xrightarrow{h} Z$. There is a canonical homotopy action

$$a\colon \Omega Z \times E \longrightarrow E$$

which extends the map $\Omega Z \vee E \xrightarrow{\partial \vee 1} E$. The composite $\Omega Y \times E \xrightarrow{\Omega h \times 1} \Omega Z \times E \xrightarrow{a} E$ therefore has the property that its restriction to ΩY is null homotopic, resulting in a quotient map

$$\Theta\colon \Omega Y \ltimes E \longrightarrow E.$$

Second, it is well known that the homotopy fibre of the pinch map $Y \vee E \longrightarrow Y$ is naturally homotopy equivalent to $\Omega Y \ltimes E$. From this we obtain a homotopy fibration diagram

$$
(2.2)\qquad
\begin{array}{ccc}
\Omega Y \ltimes E & \xrightarrow{\Gamma} & E \\
\Big\downarrow & & \Big\downarrow{\scriptstyle p} \\
Y \vee E & \xrightarrow{1\perp p} & Y \\
\Big\downarrow & & \Big\downarrow{\scriptstyle h} \\
Y & \xrightarrow{h} & Z
\end{array}
$$

for some map Γ. The link between the two constructions is the following, proved in [**Gr**].

LEMMA 2.3. *The maps Θ and Γ may be chosen so that they are homotopic.* □

Assume from now on that Θ and Γ have been chosen so that Lemma 2.3 holds. We will discuss some general properties through to Proposition 2.6, and then use the material developed to sketch a proof of Theorem 2.2. Suppose for some space B there is a map $\Sigma B \xrightarrow{\alpha} E$. One example of this will be $B = A$, where A is as in the data for Theorem 2.2 and α will be an appropriate lift for f, but other examples are also needed in Chapter 3. The naturality of the homotopy fibration $\Omega Y \ltimes E \longrightarrow Y \vee E \longrightarrow Y$ implies that there is a homotopy commutative diagram

$$\begin{array}{ccccc} \Omega Y \ltimes \Sigma B & \xrightarrow{1\ltimes\alpha} & \Omega Y \ltimes E & \xrightarrow{\Gamma} & E \\ \downarrow & & \downarrow & & \downarrow{\scriptstyle p} \\ Y \vee \Sigma B & \xrightarrow{1\vee\alpha} & Y \vee E & \xrightarrow{1\perp p} & Y. \end{array} \tag{2.3}$$

By Lemma 2.1 and its use in defining the map i, there is a natural homotopy equivalence

$$(\Omega Y \wedge \Sigma B) \vee \Sigma B \xrightarrow{i\perp j} \Omega Y \ltimes \Sigma B.$$

The composite $\Sigma B \xrightarrow{j} \Omega Y \ltimes \Sigma B \longrightarrow Y \vee \Sigma B$ is the inclusion i_2 of the second wedge summand. The composite $\Omega Y \wedge \Sigma B \xrightarrow{i} \Omega Y \ltimes \Sigma B \longrightarrow Y \vee \Sigma B$ can also be identified; it is a certain Whitehead product.

To motivate the appearance of Whitehead products, in general suppose that X_1 and X_2 are pointed, path-connected spaces. For $j = 1, 2$, let

$$i_j \colon X_j \longrightarrow X_1 \vee X_2$$

be the inclusion of the jth-wedge summand, and let ev_j be the composite

$$ev_j \colon \Sigma\Omega X_j \xrightarrow{ev} X_j \xrightarrow{i_j} X_1 \vee X_2.$$

Ganea [**Ga**] showed that there is a homotopy fibration

$$\Sigma\Omega X_1 \wedge \Omega X_2 \xrightarrow{[ev_1, ev_2]} X_1 \vee X_2 \longrightarrow X_1 \times X_2$$

where the right map is the inclusion of the wedge into the product and the left map is the Whitehead product of ev_1 and ev_2. Consider the composite

$$\Sigma\Omega X_1 \wedge \Omega X_2 \xrightarrow{[ev_1, ev_2]} X_1 \vee X_2 \xrightarrow{q_1} X_1$$

where q_1 is the pinch map to the first wedge summand. The naturality of the Whitehead product implies that this composite is homotopic to $[q_1 \circ ev_1, q_1 \circ ev_2]$. But $q_1 \circ ev_2 = q_1 \circ i_2 \circ ev$ is null homotopic since $q_1 \circ i_2$ is. Therefore $q_1 \circ [ev_1, ev_2]$ is null homotopic, so there is a lift

$$\begin{array}{ccc} & & \Omega X_1 \ltimes X_2 \\ & \overset{\xi}{\nearrow} & \downarrow \\ \Sigma\Omega X_1 \wedge \Omega X_2 & \xrightarrow{[ev_1, ev_2]} & X_1 \vee X_2 \end{array}$$

for some map ξ. Suppose that X_2 is a suspension, $X_2 \simeq \Sigma X_2'$, and let $E \colon X_2' \longrightarrow \Omega\Sigma X_2'$ be the suspension map, which is adjoint to the identity map on $\Sigma X_2'$. Then we may precompose $[ev_1, ev_2]$ with $\Sigma\Omega X_1 \wedge X_2' \xrightarrow{\Sigma 1\wedge E} \Sigma\Omega X_1 \wedge \Omega\Sigma X_2'$. The naturality of the Whitehead product implies that $[ev_1, ev_2] \circ (\Sigma 1 \wedge E) \simeq [ev_1, ev_2 \circ \Sigma E]$. Since

ΣE is a left homotopy inverse for ev, we have $ev_2 \circ \Sigma E = i_2 \circ ev \circ \Sigma E \simeq i_2$. Combining this with the lift ξ gives a homotopy commutative diagram

$$\begin{array}{ccc} & & \Omega X_1 \ltimes \Sigma X_2' \\ & \overset{\xi'}{\nearrow} & \downarrow \\ \Sigma\Omega X_1 \wedge X_2' & \xrightarrow{[ev_1, i_2]} & X_1 \vee \Sigma X_2' \end{array}$$

where $\xi' = \xi \circ (\Sigma 1 \wedge E)$. Writing $\Sigma\Omega X_1 \wedge X_2'$ as $\Omega X_1 \wedge \Sigma X_2'$, in [**BT2**] it is shown that ξ can be chosen so that ξ' is the map $\Omega X_1 \wedge \Sigma X_2' \xrightarrow{i} \Omega X_2 \ltimes \Sigma X_2'$.

Thus, returning to the case of $Y \vee \Sigma B$, there is a homotopy commutative diagram

$$\begin{array}{ccc} (\Omega Y \wedge \Sigma B) \vee \Sigma B & \xrightarrow{i \perp j} & \Omega Y \ltimes \Sigma B \\ & \searrow^{[ev_1, i_2] \perp i_2} & \downarrow \\ & & Y \vee \Sigma B \end{array} \tag{2.4}$$

and the map $i \perp j$ along the top row is a homotopy equivalence. Combining (2.3) and (2.4) and using the naturality of the Whitehead product results in a homotopy commutative diagram

$$\begin{array}{ccc} (\Omega Y \wedge \Sigma B) \vee \Sigma B & \xrightarrow{\Psi} & E \\ & \searrow^{[ev, p \circ \alpha] \perp (p \circ \alpha)} & \downarrow p \\ & & Y \end{array} \tag{2.5}$$

where $\Psi = \Gamma \circ (1 \ltimes \alpha) \circ (i \perp j)$.

Next, suppose that the map $Y \xrightarrow{h} Z$ in Theorem 2.2 has the additional property that Ωh has a right homotopy inverse $s \colon \Omega Z \longrightarrow \Omega Y$. Then the fibration connecting map $\Omega Z \longrightarrow E$ is null homotopic, so the homotopy action $\Omega Z \times E \xrightarrow{a} E$ factors as a composite

$$\Omega Z \times E \xrightarrow{\pi} \Omega Z \ltimes E \longrightarrow E$$

where π is the quotient map and the right map is a choice of extension. The next lemma shows that an extension may be chosen to be the composite

$$\bar{a} \colon \Omega Z \ltimes E \xrightarrow{s \ltimes 1} \Omega Y \ltimes E \xrightarrow{\Gamma} E.$$

LEMMA 2.4. *Suppose that the homotopy fibration $E \longrightarrow Y \xrightarrow{h} Z$ has the property that Ωh has a right homotopy inverse $s \colon \Omega Z \longrightarrow \Omega Y$. Then there is a homotopy commutative diagram*

$$\begin{array}{ccc} \Omega Z \times E & \xrightarrow{a} & E \\ \downarrow \pi & & \Vert \\ \Omega Z \ltimes E & \xrightarrow{\bar{a}} & E. \end{array}$$

PROOF. Consider the diagram

$$\begin{array}{ccccccc}
\Omega Z\times E & \xrightarrow{s\times 1} & \Omega Y\times E & \xrightarrow{\Omega h\times 1} & \Omega Z\times E & \xrightarrow{a} & E\\
\downarrow{\scriptstyle\pi} & & \downarrow{\scriptstyle\pi} & & & & \Vert\\
\Omega Z\ltimes E & \xrightarrow{s\ltimes 1} & \Omega Y\ltimes E & & \xrightarrow{\qquad\Theta\qquad} & & E.
\end{array}$$

The left square commutes by the naturality of π while the right square commutes by definition of Θ. Since s is a right homotopy inverse of Ωh, the top row is homotopic to a. Thus the homotopy commutativity of the diagram implies that $a\simeq\Theta\circ(s\ltimes 1)\circ\pi$. By Lemma 2.3, $\Theta\circ\Gamma$, and by definition, $\overline{a}=\Gamma\circ(s\ltimes 1)$. Therefore Therefore $a\simeq\Gamma\circ(s\ltimes 1)\circ\pi=\overline{a}\circ\pi$, giving the asserted homotopy commutative diagram. □

Define ϑ by the composite

$$\vartheta\colon \Omega Z\ltimes\Sigma B\xrightarrow{s\ltimes\alpha}\Omega Y\ltimes E\xrightarrow{\Gamma}E.$$

Note that as $(s\ltimes\alpha)=(1\ltimes\alpha)\circ(s\ltimes 1)$, the definitions of θ and $\overline{a}$ immediately imply the following.

LEMMA 2.5. *Suppose that the homotopy fibration* $E\longrightarrow Y\xrightarrow{h}Z$ *has the property that* Ωh *has a right homotopy inverse* $s\colon\Omega Z\longrightarrow\Omega Y$. *Then for any map* $\Sigma B\xrightarrow{\alpha}E$ *there is a homotopy commutative diagram*

$$\begin{array}{ccc}
\Omega Z\ltimes\Sigma B & & \\
\downarrow{\scriptstyle 1\ltimes\alpha} & \searrow{\scriptstyle\vartheta} & \\
\Omega Z\ltimes E & \xrightarrow{\overline{a}} & E.
\end{array}$$

□

The map ϑ is the bridge between the homotopy action, in the form of $\overline{a}$, and Whitehead products as in (2.5).

PROPOSITION 2.6. *Suppose that there is a homotopy fibration sequence* $\Omega Z\xrightarrow{\partial}E\xrightarrow{p}Y\xrightarrow{h}Z$ *where* Ωh *has a right homotopy inverse* $s\colon\Omega Z\longrightarrow\Omega Y$. *Let* $\alpha\colon\Sigma B\longrightarrow E$ *be a map. Then there is a homotopy commutative diagram*

$$\begin{array}{ccccc}
(\Omega Z\wedge\Sigma B)\vee\Sigma B & \xrightarrow{i\perp j} & \Omega Z\ltimes\Sigma B & \xrightarrow{\vartheta} & E\\
& & \searrow{\scriptstyle[ev\circ\Sigma s,\,p\circ\alpha]\perp p\circ\alpha} & & \downarrow{\scriptstyle p}\\
& & & & Y.
\end{array}$$

PROOF. Consider the diagram

$$\begin{array}{ccccccc}
(\Omega Z\wedge\Sigma B)\vee\Sigma B & \xrightarrow{(s\wedge 1)\vee 1} & (\Omega Y\wedge\Sigma B)\vee\Sigma B & & & & \\
\downarrow{\scriptstyle i\perp j} & & \downarrow{\scriptstyle i\perp j} & & \searrow{\scriptstyle[ev,p\circ\alpha]\perp(p\circ\alpha)} & & \\
\Omega Z\ltimes\Sigma B & \xrightarrow{s\ltimes 1} & \Omega Y\ltimes\Sigma B & \xrightarrow{\Gamma\circ(1\ltimes\alpha)} & E & \xrightarrow{p} & Y.
\end{array}$$

The left square homotopy commutes by the naturality of i and j while the right triangle homotopy commutes by (2.5). The composite $\Gamma\circ(1\ltimes\alpha)\circ(s\ltimes 1)=\Gamma\circ(s\ltimes\alpha)$ along the bottom row is the definition of ϑ. The naturality of the Whitehead product implies that the composite in the upper direction around the diagram is

$[ev \circ \Sigma s, p \circ \alpha] \perp (p \circ \alpha)$. Thus the homotopy commutativity of the diagram implies that $p \circ \theta \circ (i \perp j) \simeq [ev \circ \Sigma s, p \circ \alpha] \perp (p \circ \alpha)$, as asserted. $\square$

Finally, we justify Theorem 2.2. In the data for Theorem 2.2, focus on the map $\Sigma A \xrightarrow{f} Y$. The extension of h to h' implies that $h \circ f$ is null homotopic. Thus f lifts to a map $g\colon \Sigma A \longrightarrow E$. However, not every choice of lift g will also make part (a) of Theorem 2.2 hold. For part (a) to hold, the composite $\Sigma A \xrightarrow{g} E \longrightarrow E'$ must be null homotopic, or equivalently, g must factor through the homotopy fibre F of $E \longrightarrow E'$. In terms of the data in (2.1), as the upper square is a homotopy pullback, F is also the homotopy fibre of the map $Y \longrightarrow Y'$. Since Y' is the homotopy cofibre of $\Sigma A \xrightarrow{f} Y$, the map f lifts to F so for any such lift we may choose g to be the composite $\Sigma A \longrightarrow F \longrightarrow E$. Assume from now on that such a g has been chosen. Then [**BT2**] ensures that Theorem 2.2 (a) holds.

Define θ by the composite

$$\theta\colon \Omega Z \ltimes \Sigma A \xrightarrow{s \ltimes g} \Omega Y \ltimes E \xrightarrow{\Gamma} E.$$

Then θ is a special case of the map ϑ in Lemma 2.5, and as in that lemma, there is a homotopy commutative diagram

$$\begin{array}{ccc} \Omega Z \ltimes \Sigma A & & \\ \big\downarrow{\scriptstyle 1 \ltimes g} & \overset{\theta}{\searrow} & \\ \Omega Z \ltimes E & \xrightarrow{\bar{a}} & E. \end{array} \tag{2.6}$$

Applying Proposition 2.6 to g and θ we obtain a homotopy commutative diagram

$$\begin{array}{ccccc} (\Omega Z \wedge \Sigma A) \vee \Sigma A & \xrightarrow{i \perp j} & \Omega Z \ltimes \Sigma A & \xrightarrow{\theta} & E \\ & & \overset{[ev \circ \Sigma s, p \circ g] \perp p \circ g}{\searrow} & & \big\downarrow{\scriptstyle p} \\ & & & & Y, \end{array} \tag{2.7}$$

which is exactly the statement of Theorem 2.2 (b).

REMARK 2.7. Theorem 2.2 also has a naturality property, not explicitly stated in [**BT2**]. Suppose that there is a map of principal fibrations

$$\begin{array}{ccccc} \Omega Z & \longrightarrow & E & \xrightarrow{p} & Y \\ \big\downarrow & & \big\downarrow & & \big\downarrow \\ \Omega \widehat{Z} & \longrightarrow & \widehat{Z} & \xrightarrow{\widehat{p}} & \widehat{Y}. \end{array}$$

The map Γ is natural for maps of principal fibrations by [**BT2**, Proposition 2.9] and therefore, by its definition, so is $\bar{a}$ provided there is a homotopy commutative diagram of right homotopy inverses

$$\begin{array}{ccc} \Omega Z & \xrightarrow{s} & \Omega Y \\ \big\downarrow & & \big\downarrow \\ \Omega \widehat{Z} & \xrightarrow{\widehat{s}} & \Omega \widehat{Y}. \end{array} \tag{2.8}$$

The construction of the lift g in [**BT2**] is natural and hence so is the homotopy commutative diagram (2.6). The naturality of the Whitehead product then implies that (2.7) is natural. Thus the homotopy commutative diagram in Theorem 2.2 (b) is natural for maps of principal fibrations with compatible right homotopy inverses. Further, the naturality of $\theta \simeq \overline{a} \circ (1 \ltimes g)$ implies that the homotopy cofibration in Theorem 2.2 (a) is natural in the sense that we obtain a homotopy cofibration diagram

$$\begin{array}{ccccc}
\Omega Z \ltimes A & \xrightarrow{\theta} & E & \longrightarrow & E' \\
\downarrow & & \downarrow & & \downarrow \\
\Omega \widehat{Z} \ltimes \widehat{A} & \xrightarrow{\widehat{\theta}} & \widehat{E} & \longrightarrow & \widehat{E}'
\end{array}$$

where the left square homotopy commutes by the naturality of θ and the dashed arrow is an induced map of homotopy cofibres that makes the right square homotopy commute.

In conclusion, Theorem 2.2 is natural for maps of principal homotopy fibrations with the property that the right homotopy inverses satisfy the homotopy commutative diagram 2.8.

CHAPTER 3

The fibre of the pinch map

In this section we prove Theorem 1.1. This begins with some information on homotopy actions and half-smashes. In Lemma 3.2 it is shown that the homotopy associativity of the homotopy action $\Omega Z \times E \xrightarrow{a} E$ has a partial analogue in the half-smash case with respect to the map $\Omega Z \ltimes E \xrightarrow{\overline{a}} E$.

LEMMA 3.1. *Let B, C and D be pointed, path-connected spaces. Then there is a natural homeomorphism*

$$(B \times C) \ltimes D \xrightarrow{\varphi} B \ltimes (C \ltimes D)$$

satisfying commutative diagrams

$$\begin{array}{ccc} B \times C \times D & \xrightarrow{1 \times \pi} & B \times (C \ltimes D) \\ \downarrow{\scriptstyle \pi} & & \downarrow{\scriptstyle \pi} \\ (B \times C) \ltimes D & \xrightarrow{\varphi} & B \ltimes (C \ltimes D). \end{array}$$

and

$$\begin{array}{ccc} (B \times C) \ltimes D & \xrightarrow{\varphi} & B \ltimes (C \ltimes D) \\ \downarrow & & \downarrow \\ B \wedge C \wedge D & = & B \wedge C \wedge D \end{array}$$

where in the second diagram the vertical maps are the quotients to the smash products.

PROOF. The map $B \times C \times D \xrightarrow{\pi} (B \times C) \ltimes D$ identifies the subspace $B \times C \times * \subseteq B \times C \times D$ to the basepoint. On the other hand, the map $B \times (C \ltimes D) \xrightarrow{\pi} B \ltimes (C \ltimes D)$ identifies the subspace $B \times *' \subset B \times (C \ltimes D)$ to the basepoint, where $*'$ is the basepoint of $C \ltimes D$. But $*'$ is the result of identifying the subspace $C \times * \subseteq C \times D$ to the basepoint. Thus $\pi \circ (1 \times \pi)$ identifies the subspace $B \times C \times * \subseteq B \times C \times D$ to the basepoint. Therefore $(B \times C) \ltimes D$ and $B \ltimes (C \ltimes D)$ are identical as quotient spaces of $B \times C \times D$. This identification is natural since each quotient map involved is natural.

The identification of $(B \times C) \ltimes D$ and $B \ltimes (C \ltimes D)$ as identical quotient spaces of $B \times C \times D$ implies that the further quotient maps in both cases to the smash product $B \wedge C \wedge D$ are also identical, giving the second asserted commutative diagram. $\square$

Given a homotopy fibration sequence $\Omega Z \xrightarrow{\partial} E \xrightarrow{p} Y \xrightarrow{h} Z$, one property of the homotopy action $\Omega Z \times E \xrightarrow{a} E$ is that it satisfies a homotopy commutative diagram

$$
\begin{array}{ccc}
\Omega Z \times \Omega Z \times E & \xrightarrow{\mu \times 1} & \Omega Z \times E \\
\Big\downarrow{\scriptstyle 1 \times a} & & \Big\downarrow{\scriptstyle a} \\
\Omega Z \times E & \xrightarrow{\quad a \quad} & E
\end{array}
\tag{3.1}
$$

where μ is the loop space multiplication. If Ωh has a right homotopy inverse then a factors as the composite

$$\Omega Z \times E \xrightarrow{\pi} \Omega Z \ltimes E \xrightarrow{\overline{a}} E$$

where we may take $\overline{a}$ as in Lemma 2.4. Ideally, we would like (3.1) to descend to a homotopy commutative diagram

$$
\begin{array}{ccccc}
(\Omega Z \times \Omega Z) \ltimes E & & \xrightarrow{\qquad \mu \ltimes 1 \qquad} & & \Omega Z \ltimes E \\
\Big\downarrow{\scriptstyle \varphi} & & & & \Big\downarrow{\scriptstyle \overline{a}} \\
\Omega Z \ltimes (\Omega Z \ltimes E) & \xrightarrow{1 \ltimes \overline{a}} & \Omega Z \ltimes E & \xrightarrow{\quad \overline{a} \quad} & E.
\end{array}
$$

However, it is not clear if this diagram does in fact homotopy commute, as will be explained momentarily. As it would be useful to have such a diagram in what is to come, we will discuss to what extent valid information can be extracted from the diagram.

The issue is a general one. Let B, C and W be pointed, path-connected spaces. The homotopy cofibration sequence $B \longrightarrow B \times C \xrightarrow{\pi} B \ltimes C \xrightarrow{\delta} \Sigma B$ induces an exact sequence of pointed sets

$$[\Sigma B, W] \xrightarrow{\delta^*} [B \ltimes C, W] \xrightarrow{\pi^*} [B \times C, W] \longrightarrow [B, W]. \tag{3.2}$$

The inclusion of B into $B \times C$ has a left inverse given by the projection $B \times C \longrightarrow B$. Therefore δ is null homotopic, implying that δ^* is the zero map. However, as (3.2) is only an exact sequence of pointed sets the fact that $\delta^* = 0$ does not in general imply that π^* is a monomorphism. More precisely, the homotopy coaction $B \ltimes C \xrightarrow{\psi} (B \ltimes C) \vee \Sigma B$ induces an action $[B \ltimes C, W] \times [\Sigma B, W] \longrightarrow [B \ltimes C, W]$ of the group $[\Sigma B, W]$ on the set $[B \ltimes C, W]$. The orbits of this action have the property that if $a, b \in [B \ltimes C, W]$ are in different orbits then $\pi^*(a) \neq \pi^*(b)$. However, it may be the case that distinct homotopy classes are in the same orbit and both are sent by π^* to the same element. The fact that $\delta^* = 0$ only implies that the orbit of the trivial map consists of a single homotopy class.

In our case, we may not be able to use the fact that $a \circ (1 \times a) \simeq a \circ (\mu \times 1)$ to show that the quotient map $(\Omega Z \times \Omega Z) \times E \xrightarrow{\pi} (\Omega Z \times \Omega Z) \ltimes E$ induces a homotopy $\overline{a} \circ (1 \times \overline{a}) \circ \varphi \simeq \overline{a} \circ (\mu \ltimes 1)$. However, we are able to prove the following, which will suffice for our purposes.

LEMMA 3.2. *Suppose that there is a homotopy fibration sequence* $\Omega Z \xrightarrow{\partial} E \longrightarrow Y \xrightarrow{h} Z$ *where* Ωh *has a right homotopy inverse. Then the diagram*

$$\begin{array}{ccccc} (\Omega Z\times\Omega Z)\ltimes E & & \xrightarrow{\mu\ltimes 1} & & \Omega Z\ltimes E \\ \downarrow{\scriptstyle\varphi} & & & & \downarrow{\scriptstyle\overline{a}} \\ \Omega Z\ltimes(\Omega Z\ltimes E) & \xrightarrow{1\ltimes\overline{a}} & \Omega Z\ltimes E & \xrightarrow{\overline{a}} & E \end{array}$$

has the following properties:

(a) *it homotopy commutes when precomposed with* $(\Omega Z\times\Omega Z)\times E \xrightarrow{\pi} (\Omega Z\times\Omega Z)\ltimes E$;
(b) *it homotopy commutes after suspending;*
(c) *it commutes in homology.*

PROOF. First consider the diagram

$$\begin{array}{ccccc} \Omega Z\times\Omega Z\times E & \xrightarrow{\mu\times 1} & \Omega Z\times E & & \\ \downarrow{\scriptstyle\pi} & & \downarrow{\scriptstyle\pi} & \searrow{\scriptstyle a} & \\ (\Omega Z\times\Omega Z)\ltimes E & \xrightarrow{\mu\ltimes 1} & \Omega Z\ltimes E & \xrightarrow{\overline{a}} & E. \end{array}$$

The left square commutes by the naturality of the quotient map π and the right side commutes by Lemma 2.4. Thus the homotopy commutativity of the diagram implies that $a\circ(\mu\times 1)\simeq\overline{a}\circ(\mu\ltimes 1)\circ\pi$. Next consider the diagram

$$\begin{array}{ccccccc} \Omega Z\times\Omega Z\times E & \xrightarrow{1\times\pi} & \Omega Z\times(\Omega Z\ltimes E) & \xrightarrow{1\times\overline{a}} & \Omega Z\times E & & \\ \downarrow{\scriptstyle\pi} & & \downarrow{\scriptstyle\pi} & & \downarrow{\scriptstyle\pi} & \searrow{\scriptstyle a} & \\ (\Omega Z\times\Omega Z)\ltimes E & \xrightarrow{\varphi} & \Omega Z\ltimes(\Omega Z\ltimes E) & \xrightarrow{1\ltimes\overline{a}} & \Omega Z\ltimes E & \xrightarrow{\overline{a}} & E. \end{array}$$

The left square commutes by Lemma 3.1, the middle square commutes by the naturality of the quotient map π, and the right triangle commutes by Lemma 2.4. Notice that Lemma 2.4 also implies that the top row is homotopic to $1\times a$. Thus the homotopy commutativity of the diagram implies that $a\circ(1\times a)\simeq\overline{a}\circ(1\ltimes\overline{a})\circ\varphi\circ\pi$. Hence, from the property $a\circ(1\times a)\simeq a\circ(\mu\times 1)$ of a homotopy action, we obtain $\overline{a}\circ(1\ltimes\overline{a})\circ\varphi\circ\pi\simeq\overline{a}\circ(\mu\ltimes 1)\circ\pi$, proving part (a).

Since the map $\Sigma\pi$ has a right homotopy inverse, part (b) follows from part (a). Part (c) then follows from part (b) since suspending induces an isomorphism in homology. □

Lemma 3.2 motivates a definition, which appeared in a different context in [**Gr**].

DEFINITION 3.3. Let $f,g\colon X\longrightarrow Y$ be maps of pointed, path-connected spaces. Then f and g are *congruent* if $\Sigma f\simeq\Sigma g$.

Note that if f and g are homotopic then they are congruent but the converse need not hold. Note also that f congruent to g implies that, in homology, $f_*=g_*$. For our purposes, congruence is often shorthand for saying two maps induces the same map in homology. This will often be used in the context of homotopy equivalences when both X and Y are simply-connected. In this case, if f is a homotopy equivalence then it induces an isomorphism in homology, so g also induces an isomorphism in homology by the congruence property, and hence is also a homotopy equivalence by Whitehead's Theorem. Other properties of congruent maps are discussed in [**Gr**] but they will not be used here.

REMARK 3.4. Lemma 3.2 may now be rephrased as saying the composites $\overline{a}\circ(\mu\ltimes 1)$ and $\overline{a}\circ(1\ltimes\overline{a})\circ\varphi$ are congruent.

Now specialize to the homotopy fibration sequence

$$\Omega\Sigma X\xrightarrow{\partial}E\xrightarrow{p}\Sigma X\vee\Sigma Y\xrightarrow{q_1}\Sigma X$$

where q_1 is the pinch map to the first wedge summand. It is known (see, for example, [**N4**, 4.3.2]) that there is a homotopy equivalence

$$E\simeq\bigvee_{k=0}^{\infty}X^{\wedge k}\wedge\Sigma Y$$

where $X^0\wedge\Sigma Y$ is regarded as ΣY, and the map from p can be identified in terms of iterated Whitehead products based on the inclusions of ΣX and ΣY into $\Sigma X\vee\Sigma Y$. We show in Theorem 3.15 that such an equivalence can be chosen so that it also inherits properties of the homotopy action of $\Omega\Sigma X$ on E.

This begins with an initial homotopy equivalence for E that depends on a special case of Theorem 2.2 proved in [**BT2**].

THEOREM 3.5. *Suppose that there is a homotopy cofibration* $\Sigma A\xrightarrow{f}Y\xrightarrow{h}Y'$. *Let E be the homotopy fibre of h and let* $g\colon\Sigma A\longrightarrow E$ *be a lift of f. If Ωh has a right homotopy inverse* $s\colon\Omega Y'\longrightarrow\Omega Y$ *then the the lift g may be chosen so that:*

(a) *the composite* $\Omega Y'\ltimes\Sigma A\xrightarrow{1\ltimes g}\Omega Y'\ltimes E\xrightarrow{\overline{a}}E$ *is a homotopy equivalence;*
(b) *there is a homotopy fibration*

$$\Omega Y'\ltimes\Sigma A\xrightarrow{\chi}Y\xrightarrow{h}Y'$$

where χ is the sum of the two maps $\Omega Y'\ltimes\Sigma A\xrightarrow{\pi}\Sigma A\xrightarrow{f}Y$ *and* $\Omega Y'\ltimes\Sigma A\xrightarrow{q}\Omega Y'\wedge\Sigma A\xrightarrow{[evos,f]}Y$.

PROOF. While proved in [**BT2**], the proof is included to make the assertions transparent. Taking $Z=Y'$ in Theorem 2.2 gives a diagram of data

$$\begin{array}{ccccc}
 & & E & \longrightarrow & E' \\
 & & \big\downarrow{\scriptstyle p} & & \big\downarrow{\scriptstyle p'} \\
\Sigma A & \xrightarrow{f} & Y & \xrightarrow{h} & Y' \\
 & & \big\downarrow{\scriptstyle h} & & \big\| \\
 & & Y' & = & Y'.
\end{array}$$

where the vertical columns and the maps between them form a homotopy fibration diagram. Since Ωh has a right homotopy inverse, by Theorem 2.2 there is a homotopy cofibration

$$\Omega Y'\ltimes\Sigma A\xrightarrow{\theta}E\longrightarrow E'.$$

As E' is contractible, θ is a homotopy equivalence. By (2.6), θ is homotopic to the composite $\overline{a}\circ(1\ltimes g)$ for an appropriate choice of a lift g of f. This proves part (a). Defining χ as $p\circ\theta$, part (b) follows immediately from the statement of Theorem 2.2 (b). $\square$

EXAMPLE 3.6. Start with the homotopy cofibration

$$\Sigma Y \xrightarrow{i_2} \Sigma X\vee\Sigma Y \xrightarrow{q_1} \Sigma X$$

where i_2 is the inclusion of the second wedge summand. Let E be the homotopy fibre of q_1 and let $g\colon \Sigma Y\longrightarrow E$ be a lift of i_2. Since the inclusion i_1 of the first wedge summand is a right homotopy inverse for q_1, Theorem 3.5 applies to show that g can be chosen so there is a homotopy equivalence

$$\Omega\Sigma X\ltimes\Sigma Y \xrightarrow{1\ltimes g} \Omega\Sigma X\ltimes E \xrightarrow{\overline{a}} E$$

and there is a homotopy fibration

$$\Omega\Sigma X\ltimes\Sigma Y \xrightarrow{\chi} \Sigma X\vee\Sigma Y \xrightarrow{q_1} \Sigma X$$

where χ is the wedge sum of the maps $\Omega\Sigma X\ltimes\Sigma Y\xrightarrow{\pi}\Sigma Y\xrightarrow{i_2}\Sigma X\vee\Sigma Y$ and $\Omega\Sigma X\ltimes\Sigma Y\xrightarrow{\quad q\quad}\Omega\Sigma X\wedge\Sigma Y\xrightarrow{[ev\circ\Omega i_1, i_2]}\Sigma X\vee\Sigma Y$.

REMARK 3.7. Example 3.6 works equally well with ΣX replaced by a space X' that is not necessarily a suspension, but the use of ΣX is to align with later examples and applications.

Next, we build towards Theorem 3.15. In general, the James construction gives a homotopy equivalence $\Sigma\Omega\Sigma X\simeq\bigvee_{k=1}^{\infty}\Sigma X^{\wedge k}$ which is natural for maps $X\longrightarrow Y$. There are different choices of such an equivalence and it will help if we fix one. Focus on the suspension $X\xrightarrow{E}\Omega\Sigma X$. For $k\geq 1$, let e_k be the composite

$$e_k\colon X^{\times k}\xrightarrow{E^{\times k}}(\Omega\Sigma X)^{\times k}\xrightarrow{\mu}\Omega\Sigma X$$

where μ is the standard loop multiplication. There is a natural homotopy equivalence $\Sigma(A\times B)\simeq\Sigma A\vee\Sigma B\vee(\Sigma A\wedge B)$. Iterating this we obtain a natural map $\Sigma X_1\wedge\cdots\wedge X_k\longrightarrow\Sigma(X_1\times\cdots\times X_k)$. Let ϕ_k be the composite

$$\phi_k\colon \Sigma X^{\wedge k}\longrightarrow\Sigma(X^{\times k})\xrightarrow{\Sigma e_k}\Sigma\Omega\Sigma X.$$

Let

$$\phi\colon \bigvee_{k=1}^{\infty}\Sigma X^{\wedge k}\longrightarrow\Sigma\Omega\Sigma X$$

be the wedge sum of the maps ϕ_k for $k\geq 1$.

LEMMA 3.8. *The map $\bigvee_{k=1}^{\infty}\Sigma X^{\wedge k}\xrightarrow{\phi}\Sigma\Omega\Sigma X$ is a homotopy equivalence that is natural for maps $X\longrightarrow Y$.*

PROOF. By the Bott-Samelson Theorem there is an algebra isomorphism

$$H_*(\Omega\Sigma X;\mathbf{k})\cong T(\widetilde{H}_*(X;\mathbf{k}))$$

where $T(\ )$ is the free tensor algebra functor and $\mathbf{k}$ is a field. By construction, the map ϕ_k induces an isomorphism onto the suspension of the submodule of tensors of length k. Thus ϕ_* is an isomorphism. As this is true for homology with mod-p coefficients for any prime p and for rational coefficients, ϕ induces an isomorphism in integral homology. Thus ϕ is a homotopy equivalence by Whitehead's Theorem.

The naturality of ϕ follows by the naturality of e_k and the map $\Sigma X^{\wedge k}\longrightarrow\Sigma(X^{\times k})$. $\square$

We now specify a homotopy equivalence $\Omega\Sigma X \ltimes \Sigma Y \simeq \bigvee_{k=0}^{\infty} X^{\wedge k} \wedge \Sigma Y$ that will be needed to prove Theorem 1.1. Let

$$b_1 \colon X \wedge \Sigma Y \longrightarrow X \ltimes \Sigma Y$$

be the inclusion i. For $k \geq 2$, define

$$b_k \colon X^{\wedge k} \wedge \Sigma Y \longrightarrow (X^{\times k}) \ltimes \Sigma Y$$

recursively by the composite

$$X^{\wedge k} \wedge \Sigma Y \xrightarrow{=} X \wedge (X^{\wedge k-1} \wedge \Sigma Y) \xrightarrow{i} X \ltimes (X^{\wedge k-1} \wedge \Sigma Y)$$
$$\xrightarrow{1 \ltimes b_{k-1}} X \ltimes (X^{\times k-1} \ltimes \Sigma Y) \xrightarrow{\varphi^{-1}} (X \times X^{k-1}) \ltimes \Sigma Y \xrightarrow{=} (X^{\times k}) \ltimes \Sigma Y$$

where φ is the homeomorphism from Lemma 3.1.

For $k \geq 1$, let

$$q_k \colon (X^{\times k}) \ltimes \Sigma Y \longrightarrow X^{\wedge k} \wedge \Sigma Y$$

be the quotient map to the smash product.

LEMMA 3.9. *For $k \geq 1$ the composite*

$$X^{\wedge k} \wedge \Sigma Y \xrightarrow{b_k} (X^{\times k}) \ltimes \Sigma Y \xrightarrow{q_k} X^{\wedge k} \wedge \Sigma Y$$

is homotopic to the identity map.

PROOF. The proof is by induction on k. The $k = 1$ case holds because b_1 is defined as the inclusion $X \wedge \Sigma Y \xrightarrow{i} X \ltimes \Sigma Y$, the map i is a left homotopy inverse of the quotient map $X \ltimes \Sigma Y \xrightarrow{q} X \ltimes \Sigma Y$, and by definition $q_1 = q$.

For $k \geq 2$, suppose that $q_{k-1} \circ b_{k-1}$ is homotopic to the identity map. Consider the diagram

$$\begin{array}{ccccccc}
X \wedge (X^{\wedge k-1} \wedge \Sigma Y) & \xrightarrow{i} & X \ltimes (X^{\wedge k-1} \wedge \Sigma Y) & \xrightarrow{1 \ltimes b_{k-1}} & X \ltimes (X^{\times k-1} \ltimes \Sigma Y) & \xrightarrow{\varphi^{-1}} & (X \times X^{k-1}) \ltimes \Sigma Y \\
& \searrow\!\!= & \downarrow q & & \downarrow & & \downarrow \\
& & X \wedge (X^{\wedge k-1} \wedge \Sigma Y) & = & X \wedge (X^{\wedge k-1} \wedge \Sigma Y) & = & (X \wedge X^{\wedge k-1}) \wedge \Sigma Y
\end{array}$$

where the vertical maps are quotient maps to the smash product. The left triangle homotopy commutes since i is a right homotopy inverse of q, the middle square homotopy commutes by inductive hypothesis, and the right square commutes by Lemma 3.1. By definition, b_k is the top row (up to identification of $X \wedge (X^{\wedge k-1} \wedge \Sigma Y)$ as $X^{\wedge k} \wedge \Sigma Y$ and $(X \times X^{k-1}) \ltimes \Sigma Y$ as $(X^{\times k}) \ltimes \Sigma Y$) and the right vertical map can be identified with q_k. Thus $q_k \circ b_k$ is homotopic to the identity map, proving the inductive step. $\square$

Let

$$c_0 \colon \Sigma Y \longrightarrow \Omega\Sigma X \ltimes \Sigma Y$$

be the inclusion j and for $k \geq 1$ define c_k by the composite

$$c_k \colon X^{\wedge k} \wedge \Sigma Y \xrightarrow{b_k} (X^{\times k}) \ltimes \Sigma Y \xrightarrow{e_k \ltimes 1} \Omega\Sigma X \ltimes \Sigma Y.$$

Let

$$c \colon \bigvee_{k=0}^{\infty} X^{\wedge k} \wedge \Sigma Y \longrightarrow \Omega\Sigma X \ltimes \Sigma Y$$

be the wedge sum of the maps c_k for $k \geq 0$, where $X^{\wedge 0} \wedge \Sigma Y$ is understood to be ΣY.

LEMMA 3.10. *The map* $\bigvee_{k=0}^{\infty} X^{\wedge k} \wedge \Sigma Y \xrightarrow{c} \Omega\Sigma X \ltimes \Sigma Y$ *is a homotopy equivalence.*

PROOF. Take homology with field coefficients. By the Bott-Samelson Theorem there is an algebra isomorphism $H_*(\Omega\Sigma X) \cong T(V)$ where $V = \widetilde{H}_*(X)$. The homotopy equivalence $\Omega\Sigma X \ltimes \Sigma Y \simeq (\Omega\Sigma X \wedge \Sigma Y) \vee \Sigma Y$ therefore implies that there is a module isomorphism

$$H_*(\Omega\Sigma X \ltimes \Sigma Y) \cong \Big(T(V) \otimes \widetilde{H}_*(\Sigma Y)\Big) \oplus H_*(\Sigma Y).$$

By Lemma 3.9, the map $X^{\wedge k} \wedge \Sigma Y \xrightarrow{b_k} (X^{\times k}) \ltimes \Sigma Y$ is a left homotopy inverse for the quotient map $(X^{\times k}) \ltimes \Sigma Y \xrightarrow{q_k} X^{\wedge k} \wedge \Sigma Y$. Therefore the image of $(b_k)_*$ is isomorphic to the submodule $\widetilde{H}_*(X)^{\otimes k} \otimes \widetilde{H}_*(\Sigma Y)$. The map $(e_k \ltimes 1)_*$ maps this submodule isomorphically onto $V^{\otimes k} \otimes \widetilde{H}_*(\Sigma Y)$. Thus, if $k \geq 1$, as c_k is defined as $(e_k \ltimes 1) \circ b_k$, the image of $(c_k)_*$ is isomorphic to the submodule $V^{\otimes k} \otimes \widetilde{H}_*(\Sigma Y)$. Since c_0 is the inclusion of ΣY into $\Omega\Sigma X \ltimes \Sigma Y$, its image in homology is $H_*(\Sigma Y)$. Thus c_* is an isomorphism. As this is true for homology with any field coefficients, c induces an isomorphism in integral homology so c is a homotopy equivalence by Whitehead's Theorem. □

We now define two maps $\bigvee_{k=0}^{\infty} X^{\wedge k} \wedge \Sigma Y \longrightarrow E$, both of which will be homotopy equivalences, and which are congruent. First, let

$$d_0 \colon \Sigma Y \longrightarrow E$$

be g. For $k \geq 1$, let d_k be the composite

$$d_k \colon X^{\wedge k} \wedge \Sigma Y \xrightarrow{c_k} \Omega\Sigma X \ltimes \Sigma Y \xrightarrow{1 \ltimes g} \Omega\Sigma X \ltimes E \xrightarrow{\overline{a}} E.$$

Let

$$d \colon \bigvee_{k=0}^{\infty} X^{\wedge k} \wedge \Sigma Y \longrightarrow E$$

be the wedge sum of the maps d_k for $k \geq 0$. Since c is the wedge sum of the maps c_k for $k \geq 0$, the map d may equivalently be written as the composite

$$\bigvee_{k=0}^{\infty} X^{\wedge k} \wedge \Sigma Y \xrightarrow{c} \Omega\Sigma X \ltimes \Sigma Y \xrightarrow{1 \ltimes g} \Omega\Sigma X \ltimes E \xrightarrow{\overline{a}} E.$$

LEMMA 3.11. *The map d is a homotopy equivalence.*

PROOF. This follows immediately since $d = \overline{a} \circ (1 \ltimes g) \circ c$ and, by Example 3.6 and Lemma 3.10 respectively, both $\overline{a} \circ (1 \ltimes g)$ and c are homotopy equivalences. □

Next, let

$$\mathfrak{d}_0 \colon \Sigma Y \longrightarrow E$$

be g. For $k \geq 1$, let $\mathfrak{d}_k$ be the composite

$$\mathfrak{d}_k \colon X \wedge (X^{\wedge k-1} \wedge \Sigma Y) \xrightarrow{\ i\ } X \ltimes (X^{\wedge k-1} \wedge \Sigma Y) \xrightarrow{E \ltimes \mathfrak{d}_{k-1}} \Omega\Sigma X \ltimes E \xrightarrow{\ \overline{a}\ } E.$$

Let

$$\mathfrak{d} \colon \bigvee_{k=0}^{\infty} X^{\wedge k} \wedge \Sigma Y \longrightarrow E$$

be the wedge sum of the maps $\eth_k$ for $k \geq 0$. We will show that $\eth$ is congruent to d, that $\eth$ is a homotopy equivalence, and that it lifts a certain wedge sum of Whitehead products on $\Sigma X \vee \Sigma Y$.

LEMMA 3.12. *If $k = 0$ or $k = 1$ then $\eth_k = d_k$. If $k \geq 2$ then $\eth_k$ is congruent to d_k.*

PROOF. First, observe that $\eth_0 = d_0$ since, by their definitions, both equal g. Next, by definition, $\eth_1 = \overline{a} \circ (E \ltimes d_0) \circ i = \overline{a} \circ (E \ltimes g) \circ i$. On the other hand, by definition of d_1 and c_1 we have $d_1 = \overline{a} \circ (1 \ltimes g) \circ c_1 = \overline{a} \circ (1 \ltimes g) \circ (e_1 \ltimes 1) \circ b_1$. By definition, $e_1 = E$ and $b_1 = i$, so we obtain $d_1 = \overline{a} \circ (E \ltimes g) \circ i$. Thus $\eth_1 = d_1$.

Now suppose that $k \geq 2$ and assume inductively that $\eth_{k-1}$ is congruent to d_{k-1}. By definition, $d_k = \overline{a} \circ (1 \ltimes g) \circ c_k$ and $c_k = (e_k \ltimes 1) \circ b_k$, so d_k is equivalently described by the composite

$$X^{\wedge k} \wedge Y \xrightarrow{b_k} (X^{\times k}) \ltimes \Sigma Y \xrightarrow{e_k \ltimes g} \Omega\Sigma X \ltimes E \xrightarrow{\overline{a}} E. \tag{3.3}$$

Consider the diagram

$$\begin{array}{ccccccc}
X\ltimes(X^{\times k-1}\ltimes\Sigma Y) & \xrightarrow{E\ltimes(e_{k-1}\ltimes g)} & \Omega\Sigma X\ltimes(\Omega\Sigma X\ltimes E) & & \xrightarrow{1\ltimes\overline{a}} & & \Omega\Sigma X\ltimes E \\
\downarrow{\scriptstyle\varphi^{-1}} & & \downarrow{\scriptstyle\varphi^{-1}} & & & & \downarrow{\scriptstyle\overline{a}} \\
(X\times X^{\times k-1})\ltimes\Sigma Y & \xrightarrow{(E\times e_{k-1})\ltimes g} & (\Omega\Sigma X\times\Omega\Sigma X)\ltimes E & \xrightarrow{\mu\ltimes 1} & \Omega\Sigma X\ltimes E & \xrightarrow{\overline{a}} & E
\end{array}$$

where φ is the homeomorphism in Lemma 3.1. The left square commutes by the naturality of φ. The right rectangle may not homotopy commute but the two ways around the diagram are congruent by Lemma 3.2. By definition, $e_k = \mu \circ E^{\times k}$ where μ is an iterated loop multiplication on $\Omega\Sigma X$. Thus the composite

$$X \times X^{\times k-1} \xrightarrow{E\times E^{\times k-1}} \Omega\Sigma X \times (\Omega\Sigma X)^{\times k-1} \xrightarrow{1\times\mu} \Omega\Sigma X \times \Omega\Sigma X \xrightarrow{\mu} \Omega\Sigma X$$

is, on the one hand, $\mu \circ (E \times e_{k-1})$, and on the other hand, e_k. Therefore the bottom row in the diagram is $\overline{a} \circ (e_k \ltimes g)$.

Next, by definition of b_k, there is a commutative diagram

$$\begin{array}{ccccc}
X\wedge(X^{\wedge k-1}\wedge\Sigma Y) & \xrightarrow{i} & X\ltimes(X^{\wedge k-1}\wedge\Sigma Y) & \xrightarrow{1\ltimes b_{k-1}} & X\ltimes(X^{\times k-1}\ltimes\Sigma Y) \\
 & \searrow^{b_k} & & & \downarrow{\scriptstyle\varphi^{-1}} \\
 & & & & (X\times X^{\times k-1})\ltimes\Sigma Y.
\end{array}$$

Juxtapose the two previous diagrams. In the lower direction we obtain $\overline{a}\circ(e_k\ltimes g)\circ b_k$ which, by (3.3), is d_k. The upper direction is $\overline{a}\circ(1\ltimes\overline{a})\circ(E\ltimes(e_{k-1}\ltimes g))\circ(1\ltimes b_{k-1})\circ i$ which may be rewritten as $\overline{a} \circ (E \ltimes (\overline{a} \circ (e_{k-1} \ltimes g) \circ b_{k-1})) \circ i$. By (3.3) this is the same as $\overline{a} \circ (E \ltimes d_{k-1}) \circ i$. Hence d_k is congruent to $\overline{a} \circ (E \ltimes d_{k-1}) \circ i$.

By the inductive hypothesis, d_{k-1} is congruent to $\eth_{k-1}$ so $\overline{a} \circ (E \ltimes d_{k-1}) \circ i$ is congruent to $\overline{a} \circ (E \ltimes \eth_{k-1}) \circ i$. By definition, $\eth_k = \overline{a} \circ (E \ltimes \eth_{k-1}) \circ i$. Hence d_k is congruent to $\eth_k$. □

LEMMA 3.13. *The map $\eth$ is congruent to d. Consequently, $\eth$ is a homotopy equivalence.*

PROOF. By Lemma 3.12, d_k is congruent to ∂_k for all $k \geq 0$. Since d and ∂ are the wedge sums of the maps d_k and ∂_k respectively, we obtain that d is congruent to ∂. Congruence implies that $d_* = \partial_*$. By Lemma 3.11, d is a homotopy equivalence so d_* is an isomorphism. Hence ∂_* is an isomorphism and so is also a homotopy equivalence by Whitehead's Theorem. $\square$

We next show that ∂ lifts certain Whitehead products. In general, given maps $u\colon \Sigma A \longrightarrow Z$ and $v\colon \Sigma B \longrightarrow Z$ define the iterated Whitehead product

$$ad^k(u)(v)\colon A^{\wedge k} \wedge \Sigma B \longrightarrow Z$$

recursively as follows. If $k = 0$ then $ad^0(u)(v) = v$. If $k > 0$ then $ad^k(u)(v) = [u, ad^{k-1}(u)(v)]$. In our case the roles of u and v will be played by the inclusions

$$i_1\colon \Sigma X \longrightarrow \Sigma X \vee \Sigma Y \qquad i_2\colon \Sigma Y \longrightarrow \Sigma X \vee \Sigma Y$$

of the left and right wedge summands respectively.

LEMMA 3.14. *For each $k \geq 1$ there is a homotopy commutative diagram*

$$\begin{array}{ccc} X^{\wedge k} \wedge \Sigma Y & \xrightarrow{\partial_k} & E \\ & \searrow^{ad^k(i_1)(i_2)} & \downarrow p \\ & & \Sigma X \vee \Sigma Y. \end{array}$$

PROOF. The proof is by induction on k. For the base case when $k = 1$ we want to show that $p \circ \partial_1 \simeq [i_1, i_2]$. By Lemma 3.12, $\partial_1 = d_1$ so it is equivalent to show that $p \circ d_1 \simeq [i_1, i_2]$. Consider the diagram

$$(3.4)\qquad \begin{array}{ccccccc} X \wedge \Sigma Y & \xrightarrow{i} & X \ltimes \Sigma Y & & & & \\ \downarrow E\wedge 1 & & \downarrow E\ltimes 1 & & & & \\ \Omega\Sigma X \wedge \Sigma Y & \xrightarrow{i} & \Omega\Sigma X \ltimes \Sigma Y & \xrightarrow{1\ltimes g} & \Omega\Sigma X \ltimes E & \xrightarrow{\overline{a}} & E \\ & \searrow & & & & & \downarrow p \\ & & & [ev\circ\Sigma\Omega i_1, p\circ g] & & \longrightarrow & \Sigma X \vee \Sigma Y. \end{array}$$

The top left square homotopy commutes by the naturality of i. The lower triangle homotopy commutes by Proposition 2.6 with $B = Y$ and $\alpha = g$. Observe that the composite $\overline{a} \circ (1 \ltimes g) \circ (E \ltimes 1) \circ i$ along the top direction around the diagram is the definition of d_1.

Now consider the composite in the lower direction around (3.4). Write the identity map on ΣY as the suspension of the identity map on Y. So we are considering $[ev \circ \Sigma\Omega i_1, p \circ g] \circ (E \wedge \Sigma 1)$. Observe that $\Sigma X \xrightarrow{i_1} \Sigma X \vee \Sigma Y$ is a suspension, say $i_1 \simeq \Sigma i_1'$, so the naturality of E implies that $\Omega i_1 \circ E \simeq \Omega\Sigma i_1' \circ E \simeq E \circ i_1'$. As ev is a right homotopy inverse for ΣE we obtain $ev \circ \Sigma\Omega i_1 \circ \Sigma E \simeq ev \circ \Sigma E \circ \Sigma i_1' \simeq \Sigma i_1' \simeq i_1$. Therefore the naturality of the Whitehead product and the fact that $p \circ g = i_2$ imply that

$$[ev \circ \Sigma\Omega i_1, p \circ g] \circ (E \wedge \Sigma 1) \simeq [ev \circ \Sigma\Omega i_1 \circ \Sigma E, p \circ g] \simeq [i_1, i_2].$$

Thus the homotopy commutativity of (3.4) implies that $p \circ d_1 \simeq [i_1, i_2]$.

Assume inductively that $p \circ \eth_{k-1} \simeq ad^{k-1}(i_1)(i_2)$. Consider the diagram

$$\begin{array}{ccccccc}
X \wedge (X^{\wedge k-1} \wedge \Sigma Y) & \xrightarrow{i} & X \ltimes \Sigma(X^{\wedge k-1} \wedge Y) & & & & \\
\downarrow{\scriptstyle E \wedge 1} & & \downarrow{\scriptstyle E \ltimes 1} & & & & \\
\Omega\Sigma X \wedge (X^{\wedge k-1} \wedge \Sigma Y) & \xrightarrow{i} & \Omega\Sigma X \ltimes \Sigma(X^{\wedge k-1} \wedge Y) & \xrightarrow{1 \ltimes \eth_{k-1}} & \Omega\Sigma X \ltimes E & \xrightarrow{\overline{a}} & E \\
& \searrow{\scriptstyle [ev \circ \Sigma\Omega i_1, p \circ \eth_{k-1}]} & & & & & \downarrow{\scriptstyle p} \\
& & & & & & \Sigma X \vee \Sigma Y.
\end{array}$$

The top left square homotopy commutes by the naturality of i. The lower triangle homotopy commutes by Proposition 2.6 with $B = X^{\wedge k-1} \wedge Y$, $Z = \Sigma X$ and $\alpha = \eth_{k-1}$. The composite $\overline{a} \circ (1 \ltimes \eth_{k-1}) \circ (E \ltimes 1) \circ i$ along the top direction around the diagram is the definition of $\eth_k$. Again writing the identity map on ΣY as the suspension of the identity map on Y, the naturality of the Whitehead product and the inductive hypothesis $p \circ \eth_{k-1} \simeq ad^{k-1}(i_1)(i_2)$ then imply that

$$\begin{aligned}
[ev \circ \Sigma\Omega i_1, p \circ \eth_{k-1}] \circ (E \wedge \Sigma 1) &\simeq [ev \circ \Sigma\Omega i_1 \circ \Sigma E, p \circ \eth_{k-1}] \\
&\simeq [i_1, ad^{k-1}(i_1)(i_2)] \\
&= ad^k(i_1)(i_2).
\end{aligned}$$

Thus the diagram implies that $p \circ \eth_k \simeq ad^k(i_1)(i_2)$, completing the induction. □

Putting all this together we are able to prove Theorem 1.1, re-stated as follows.

THEOREM 3.15. *Let X and Y be path-connected, pointed spaces and consider the homotopy fibration $E \longrightarrow \Sigma X \vee \Sigma Y \xrightarrow{q_1} \Sigma X$. There is a homotopy commutative diagram*

$$\begin{array}{ccc}
\bigvee_{k=0}^{\infty} X^{\wedge k} \wedge \Sigma Y & \xrightarrow{\eth} & E \\
& \searrow{\scriptstyle \bigvee_{k=0}^{\infty} ad^k(i_1)(i_2)} & \downarrow{\scriptstyle p} \\
& & \Sigma X \vee \Sigma Y
\end{array}$$

where:

(a) *$\eth$ is a homotopy equivalence;*
(b) *$\eth$ is congruent to d, where d is the composite*

$$\bigvee_{k=0}^{\infty} X^{\wedge k} \wedge \Sigma Y \xrightarrow{c} \Omega\Sigma X \ltimes \Sigma Y \xrightarrow{1 \ltimes g} \Omega\Sigma X \ltimes E \xrightarrow{\overline{a}} E.$$

PROOF. By definition, $\eth$ is the wedge sum of the maps $\eth_k$ for $k \geq 0$. When $k = 0$ we have $\eth_0 = g$ where $p \circ g = i_2$ while $ad^0(i_1)(i_2) = i_2$. When $k \geq 1$, by Lemma 3.14 we have $p \circ \eth_k \simeq ad^k(i_1)(i_2)$. Thus $p \circ \eth \simeq \bigvee_{k=0}^{\infty} ad^k(i_1)(i_2)$, implying that the asserted diagram homotopy commutes.

Parts (a) and (b) are proved by Lemma 3.13. □

CHAPTER 4

Based loops on certain 2-cones

The main result in this section is Theorem 4.6, which will then be specialized to prove Theorem 1.2. We go on to give applications to Moore's conjecture.

In general, start with the data

$$\begin{CD} @. E @>>> E' \\ @. @VpVV @VVV \\ \Sigma A @>f>> Y @>>> Y' \\ @. @VhVV @VVV \\ @. Z @= Z \end{CD}$$

and suppose that Ωh has a right homotopy inverse. By Theorem 2.2 (a) there is a homotopy cofibration

$$\Omega Z \ltimes \Sigma A \xrightarrow{\theta} E \longrightarrow E'$$

where the restriction of θ to ΣA is a map

$$g\colon \Sigma A \longrightarrow E$$

which lifts f through p. The goal is to determine the homotopy type of E' by knowing properties of the space E and the map θ. In Theorem 4.6 several hypotheses are given which will let us do this.

One hypothesis is that Z is a suspension. Rewriting the data, we have

$$\begin{CD} @. E @>>> E' \\ @. @VpVV @VVV \\ \Sigma A @>f>> Y @>>> Y' \\ @. @VhVV @VVV \\ @. \Sigma X @= \Sigma X \end{CD}$$

where Ωh has a right homotopy inverse, there is a homotopy cofibration

$$\Omega\Sigma X \ltimes \Sigma A \xrightarrow{\theta} E \longrightarrow E'$$

and the restriction of θ to ΣA is a map $g\colon \Sigma A \longrightarrow E$ that lifts f through p. The appearance of $\Omega\Sigma X$ lets us take advantage of the James construction.

For $k \geq 0$, let $J_k(X)$ be the kth-stage of the James construction. Explicitly, $J_0(X) = *$ and if $k \geq 1$ then $J_k(X) = X^{\times k}/\sim$ where $(x_1, \ldots, x_t, *, x_{t+1}, \ldots, x_k) \sim (x_1, \ldots, *, x_t, x_{t+1}, \ldots, x_k)$. There is an inclusion $J_k(X) \longrightarrow J_{k+1}(X)$ given by sending $(x_1, \ldots, x_k)$ to $(x_1, \ldots, x_k, *)$. Taking a direct limit gives the space $J(X)$, and James [**J1**] showed that there is a homotopy equivalence of H-spaces $J(X) \simeq \Omega\Sigma X$. Let j_k be the composite

$$j_k\colon J_k(X) \longrightarrow J(X) \xrightarrow{\simeq} \Omega\Sigma X.$$

Let D be any pointed, path-connected space. For $k \geq 1$ let I_k be the composite

$$I_k\colon \Omega\Sigma X \ltimes (X^{\wedge k} \wedge \Sigma D) \xrightarrow{1\ltimes c_k} \Omega\Sigma X \ltimes (\Omega\Sigma X \ltimes \Sigma D) \xrightarrow{\varphi^{-1}} (\Omega\Sigma X \times \Omega\Sigma X) \ltimes \Sigma D$$

where c_k was defined in Chapter 3 and φ is the homeomorphism in Lemma 3.1. Recall that, generically, the map $B \xrightarrow{j} A \ltimes B$ is the inclusion. For $k \geq 1$ let J_k be the composite

$$J_k\colon J_{k-1}(X)\ltimes\Sigma D \xrightarrow{j_{k-1}\ltimes 1} \Omega\Sigma X \ltimes \Sigma D \xrightarrow{j} \Omega\Sigma X\ltimes(\Omega\Sigma X\ltimes\Sigma D) \xrightarrow{\varphi^{-1}} (\Omega\Sigma X\times\Omega\Sigma X)\ltimes\Sigma D.$$

LEMMA 4.1. *The composite*

$$\Omega\Sigma X\ltimes(X^{\wedge k}\wedge\Sigma D)\vee(J_{k-1}(X)\ltimes\Sigma D) \xrightarrow{I_k\perp J_k} (\Omega\Sigma X\times\Omega\Sigma X)\ltimes\Sigma D \xrightarrow{\mu\ltimes 1} \Omega\Sigma X\ltimes\Sigma D$$

is a homotopy equivalence.

PROOF. Take homology with field coefficients. Let $V = \widetilde{H}_*(X)$. By the Bott-Samelson Theorem there is an algebra isomorphism $H_*(\Omega\Sigma X) \cong T(V)$. Let us rewrite this as a module isomorphism $\widetilde{H}_*(\Omega\Sigma X) \cong \bigoplus_{t=1}^{\infty} V^{\otimes t}$. Therefore there is a module isomorphism

$$\widetilde{H}_*(\Omega\Sigma X \ltimes \Sigma D) \cong \bigoplus_{t=0}^{\infty} V^{\otimes t} \otimes \widetilde{H}_*(\Sigma D)$$

where we regard $V^0 \otimes \widetilde{H}_*(\Sigma D)$ as $\widetilde{H}_*(\Sigma D)$.

In homology, the map $J_{k-1}(X) \xrightarrow{j_{k-1}} \Omega\Sigma X$ induces the injection $\bigotimes_{t=1}^{k-1} V^{\otimes t} \longrightarrow \bigotimes_{t=1}^{\infty} V^{\otimes t}$. Observe that the composite

$$\Omega\Sigma X \ltimes \Sigma D \xrightarrow{j} \Omega\Sigma X \ltimes (\Omega\Sigma X \ltimes \Sigma D) \xrightarrow{\varphi^{-1}} (\Omega\Sigma X \times \Omega\Sigma X) \ltimes \Sigma D \xrightarrow{\mu\ltimes 1} \Omega\Sigma X \ltimes \Sigma D$$

is homotopic to the identity map: for if the domain $\Omega\Sigma X \ltimes \Sigma D$ is regarded as $* \ltimes (\Omega\Sigma X \ltimes \Sigma D)$ then j can be regarded as $b \ltimes (1 \ltimes 1)$ where b is the inclusion of the basepoint, so the naturality of φ implies that $\varphi^{-1} \circ j$ is equal to the composite $* \ltimes (\Omega\Sigma X \ltimes \Sigma D) \xrightarrow{\varphi^{-1}} (* \times \Omega\Sigma X) \ltimes \Sigma D \xrightarrow{(b\times 1)\ltimes 1} (\Omega\Sigma X \times \Omega\Sigma X) \ltimes \Sigma D$, implying that $(\mu \ltimes 1) \circ \varphi^{-1} \circ j$ is homotopic to the identity map. Therefore in homology the map $(\mu \ltimes 1) \circ J_k$ induces the same map as $j_{k-1} \ltimes 1$, which is the injection $\bigotimes_{t=1}^{k-1} V^{\otimes t} \otimes \widetilde{H}_*(\Sigma D) \longrightarrow \bigotimes_{t=1}^{\infty} V^{\otimes t} \otimes \widetilde{H}_*(\Sigma D)$. On the other hand, as in the proof of Lemma 3.10, in homology the map c_k induces the inclusion $V^{\otimes k} \otimes \widetilde{H}_*(\Sigma D) \longrightarrow \bigotimes_{t=0}^{\infty} V^{\otimes t} \otimes \widetilde{H}_*(\Sigma D)$. Therefore $(\mu \ltimes 1) \circ I_k$ induces the inclusion

$$\bigotimes_{t=k}^{\infty} V^{\otimes k} \otimes \widetilde{H}_*(\Sigma D) \longrightarrow \bigotimes_{t=0}^{\infty} V^{\otimes t} \otimes \widetilde{H}_*(\Sigma D).$$

Hence $I_k \perp J_k$ induces an isomorphism in homology. As this is true for homology with mod-p coefficients for all primes p and rational coefficients, $I_k \perp J_k$ therefore induces an isomorphism in integral homology, and so is a homotopy equivalence by Whitehead's Theorem. □

Next, suppose that there is a map

$$\delta\colon \Sigma D \longrightarrow E.$$

For $k \geq 0$, let $\overline{c}_k$ be the composite

$$\overline{c}_k\colon X^{\wedge k} \wedge \Sigma D \xrightarrow{c_k} \Omega\Sigma X \ltimes \Sigma D \xrightarrow{1\ltimes\delta} \Omega\Sigma X \ltimes E$$

and let $\overline{J}_k$ be the composite

$$\overline{J}_k\colon J_{k-1}(X) \ltimes \Sigma D \xrightarrow{j_{k-1}\ltimes\delta} \Omega\Sigma X \ltimes E \xrightarrow{\ \ j\ \ } \Omega\Sigma X \ltimes (\Omega\Sigma X \ltimes E).$$

LEMMA 4.2. *There is a homotopy commutative diagram*

$$\begin{array}{ccc}
(\Omega\Sigma X \ltimes (X^{\wedge k}\wedge\Sigma D)) \vee (J_{k-1}(X)\ltimes\Sigma D) & \xrightarrow{I_k\perp J_k} & (\Omega\Sigma X \times \Omega\Sigma X)\ltimes\Sigma D \\
 & & \big\downarrow{\scriptstyle (1\times 1)\ltimes\delta} \\
 & \searrow{\scriptstyle (1\ltimes\overline{c}_k)\perp\overline{J}_k} & (\Omega\Sigma X \times \Omega\Sigma X)\ltimes E \\
 & & \big\downarrow{\scriptstyle \varphi} \\
 & & \Omega\Sigma X \ltimes (\Omega\Sigma X \ltimes E).
\end{array}$$

PROOF. It suffices to show that the diagram homotopy commutes when restricted to each wedge summand in the domain. Observe that the definition of I_k as $\varphi^{-1}\circ(1\ltimes c_k)$ and the naturality of φ give a homotopy commutative diagram

$$\begin{array}{ccccc}
(\Omega\Sigma X \ltimes (X^{\wedge k}\wedge\Sigma D)) & \xrightarrow{I_k} & (\Omega\Sigma X\times\Omega\Sigma X)\ltimes\Sigma D & \xrightarrow{(1\times1)\ltimes\delta} & (\Omega\Sigma X\times\Omega\Sigma X)\ltimes E \\
 & \searrow{\scriptstyle 1\ltimes c_k} & \big\downarrow{\scriptstyle\varphi} & & \big\downarrow{\scriptstyle\varphi} \\
 & & \Omega\Sigma X\ltimes(\Omega\Sigma X\ltimes\Sigma D) & \xrightarrow{1\ltimes(1\ltimes\delta)} & \Omega\Sigma X\ltimes(\Omega\Sigma X\ltimes E).
\end{array}$$

By definition, $\overline{c}_k = (1\ltimes\delta)\circ c_k$ so the composite in the lower direction around the diagram is $1\ltimes\overline{c}_k$. Thus $\varphi\circ((1\times 1)\ltimes\delta)\circ I_k \simeq 1\ltimes\overline{c}_k$, as asserted.

Next, the naturality of φ and j give a homotopy commutative diagram

$$\begin{array}{ccccccc}
J_{k-1}(X)\ltimes\Sigma D & \xrightarrow{j_{k-1}\ltimes 1} & \Omega\Sigma X\ltimes\Sigma D & \xrightarrow{j} & \Omega\Sigma X\ltimes(\Omega\Sigma X\ltimes\Sigma D) & \xrightarrow{\varphi^{-1}} & (\Omega\Sigma X\times\Omega\Sigma X)\ltimes\Sigma D \\
 & \searrow{\scriptstyle j_{k-1}\ltimes\delta} & \big\downarrow{\scriptstyle 1\ltimes\delta} & & \big\downarrow{\scriptstyle 1\ltimes(1\ltimes\delta)} & & \big\downarrow{\scriptstyle (1\times1)\ltimes\delta} \\
 & & \Omega\Sigma X\ltimes E & \xrightarrow{j} & \Omega\Sigma X\ltimes(\Omega\Sigma X\ltimes E) & \xrightarrow{\varphi^{-1}} & (\Omega\Sigma X\times\Omega\Sigma X)\ltimes E.
\end{array}$$

Observe that the top row is the definition of J_k while along the bottom row the composite $j\circ(j_{k-1}\ltimes\delta)$ is the definition of $\overline{J}_k$. Therefore the diagram implies that $((1\times 1)\ltimes\delta)\circ J_k \simeq \varphi^{-1}\circ\overline{J}_k$. Thus $\varphi\circ((1\times1)\ltimes\delta)\circ J_k \simeq \varphi\circ\varphi^{-1}\circ\overline{J}_k \simeq \overline{J}_k$, as asserted. □

PROPOSITION 4.3. *Suppose that there is a homotopy fibration sequence*

$$\Omega\Sigma X \xrightarrow{\partial} E \xrightarrow{p} Y \xrightarrow{h} \Sigma X$$

where Ωh has a right homotopy inverse. Suppose that there is a map $\delta\colon \Sigma D \longrightarrow E$ such that the composite

$$\Omega\Sigma X \ltimes \Sigma D \xrightarrow{1\ltimes\delta} \Omega\Sigma X \ltimes E \xrightarrow{\overline{a}} E$$

is a homotopy equivalence. Then the composite

$$\Omega\Sigma X \ltimes (X^{\wedge k} \wedge \Sigma D) \vee (J_{k-1}(X) \ltimes \Sigma D) \xrightarrow{(1\ltimes\overline{c}_k)\perp\overline{J}_k} \Omega\Sigma X \ltimes (\Omega\Sigma X \ltimes E) \xrightarrow{1\ltimes\overline{a}} \Omega\Sigma X \ltimes E \xrightarrow{\overline{a}} E$$

is a homotopy equivalence.

PROOF. Consider the diagram

$$\begin{array}{ccccc}
(\Omega\Sigma X \ltimes (X^{\wedge k} \wedge \Sigma D)) \vee (J_{k-1}(X) \ltimes \Sigma D) & \xrightarrow{I_k\perp J_k} & (\Omega\Sigma X \times \Omega\Sigma X) \ltimes \Sigma D & \xrightarrow{\mu\ltimes 1} & \Omega\Sigma X \ltimes \Sigma D \\
 & & \Big\downarrow{\scriptstyle (1\times 1)\ltimes\delta} & & \Big\downarrow{\scriptstyle 1\ltimes\delta} \\
 & \searrow{\scriptstyle (1\ltimes\overline{c}_k)\perp\overline{J}_k} & (\Omega\Sigma X \times \Omega\Sigma X) \ltimes E & \xrightarrow{\mu\ltimes 1} & \Omega\Sigma X \ltimes E \\
 & & \Big\downarrow{\scriptstyle \varphi} & & \Big\downarrow{\scriptstyle \overline{a}} \\
 & & \Omega\Sigma X \ltimes (\Omega\Sigma X \ltimes E) \xrightarrow{1\ltimes\overline{a}} \Omega\Sigma X \ltimes E & \xrightarrow{\overline{a}} & E.
\end{array}$$

The left triangle homotopy commutes by Lemma 4.2 and the upper right square clearly commutes. The lower right square may not homotopy commute but by Lemma 3.2 the two ways around the square are congruent. The top row is a homotopy equivalence by Lemma 4.1 and the right column is a homotopy equivalence by hypothesis. Therefore the upper direction around the diagram is a homotopy equivalence. In particular, it induces an isomorphism in homology. As congruent maps induce isomorphisms in homology, this implies that the entire diagram commutes in homology, and therefore the lower direction around the diagram also induces an isomorphism in homology. Thus the lower direction around the diagram is a homotopy equivalence by Whitehead's Theorem, proving the lemma. □

Now we begin a process of altering the homotopy equivalence in Proposition 4.3 by one in which the composite $\overline{a} \circ \overline{c}_k$ has been replaced by a congruent composite $\overline{a} \circ \overline{\mathfrak{c}}_k$. As in Chapter 3, the point is that $\overline{c}_k$ behaves well with respect to multiplications, making Proposition 4.3 easy to prove, while $\overline{\mathfrak{c}}_k$ behaves well with respect to Whitehead products, which is where the applications lie.

Return to the starting map $\delta\colon \Sigma D \longrightarrow E$. Let

$$\overline{\mathfrak{c}}_0\colon \Sigma D \longrightarrow E$$

be δ. For $k \geq 1$, let $\overline{\mathfrak{c}}_k$ be the composite

$$\overline{\mathfrak{c}}_k\colon X^{\wedge k} \wedge \Sigma D \simeq X \wedge (X^{\wedge k-1} \wedge \Sigma D) \xrightarrow{\ i\ } X \ltimes (X^{\wedge k-1} \wedge \Sigma D) \xrightarrow{E\ltimes\overline{\mathfrak{c}}_{k-1}} \Omega\Sigma X \ltimes E.$$

Define $\overline{d}_0 = \overline{c}_0$ and $\overline{\mathfrak{d}}_0 = \overline{\mathfrak{c}_0}$ (so $\overline{d}_0 = \overline{\mathfrak{d}}_0 = \delta$), and for $k \geq 1$ define $\overline{d}_k$ and $\overline{\mathfrak{d}}_k$ by the composites

$$\overline{d}_k\colon X^{\wedge k} \wedge D \xrightarrow{\overline{c}_k} \Omega\Sigma X \ltimes E \xrightarrow{\overline{a}} E$$

$$\overline{\mathfrak{d}}_k\colon X^{\wedge k} \wedge D \xrightarrow{\overline{\mathfrak{c}}_k} \Omega\Sigma X \ltimes E \xrightarrow{\overline{a}} E$$

LEMMA 4.4. *If $k = 0$ or $k = 1$ then $\overline{d}_k = \overline{\mathfrak{d}}_k$. If $k \geq 2$ then $\overline{d}_k$ is congruent to $\overline{\mathfrak{d}}_k$.*

PROOF. Argue just as for Lemma 3.12, replacing $\Sigma Y \xrightarrow{g} E$ by $\Sigma D \xrightarrow{\delta} E$. □

The congruence between $\overline{d}_k$ and $\overline{\mathfrak{d}}_k$ lets us alter Proposition 4.3 to the following.

PROPOSITION 4.5. *Suppose that there is a homotopy fibration sequence*

$$\Omega\Sigma X \xrightarrow{\partial} E \xrightarrow{p} Y \xrightarrow{h} \Sigma X$$

where Ωh has a right homotopy inverse. Suppose that there is a map $\delta\colon \Sigma D \longrightarrow E$ such that the composite

$$\Omega\Sigma X \ltimes \Sigma D \xrightarrow{1\ltimes\delta} \Omega\Sigma X \ltimes E \xrightarrow{\overline{a}} E$$

is a homotopy equivalence. Then the composite

$$\Omega\Sigma X\ltimes(X^{\wedge k}\wedge\Sigma D)\vee(J_{k-1}(X)\ltimes\Sigma D) \xrightarrow{(1\ltimes\overline{\mathfrak{c}}_k)\perp\overline{J}_k} \Omega\Sigma X\ltimes(\Omega\Sigma X\ltimes E) \xrightarrow{1\ltimes\overline{a}} \Omega\Sigma X\ltimes E \xrightarrow{\overline{a}} E$$

is a homotopy equivalence.

PROOF. By Proposition 4.3 there is a homotopy equivalence
(4.1)

$$\Omega\Sigma X\ltimes(X^{\wedge k}\wedge\Sigma D)\vee(J_{k-1}(X)\ltimes\Sigma D) \xrightarrow{(1\ltimes\overline{c}_k)\perp\overline{J}_k} \Omega\Sigma X\ltimes(\Omega\Sigma X\ltimes E) \xrightarrow{1\ltimes\overline{a}} \Omega\Sigma X\ltimes E \xrightarrow{\overline{a}} E.$$

The restriction to $\Omega\Sigma X \ltimes (X^{\wedge k} \wedge \Sigma D)$ is $\overline{a}\circ(1\ltimes\overline{a})\circ(1\ltimes\overline{c}_k)$. By definition of $\overline{d}_k$, this equals $\overline{a}\circ(1\ltimes\overline{d}_k)$. By Lemma 4.4, $\overline{d}_k$ is congruent to $\overline{\mathfrak{d}}_k$, so $\overline{a}\circ(1\ltimes\overline{d}_k)$ is congruent to $\overline{a}\circ(1\ltimes\overline{\mathfrak{d}}_k)$, which by definition of $\overline{\mathfrak{d}}_k$ equals $\overline{a}\circ(1\ltimes\overline{a})\circ(1\ltimes\overline{\mathfrak{c}}_k)$. Thus the composite
(4.2)

$$\Omega\Sigma X\ltimes(X^{\wedge k}\wedge\Sigma D)\vee(J_{k-1}(X)\ltimes\Sigma D) \xrightarrow{(1\ltimes\overline{\mathfrak{c}}_k)\perp\overline{J}_k} \Omega\Sigma X\ltimes(\Omega\Sigma X\ltimes E) \xrightarrow{1\ltimes\overline{a}} \Omega\Sigma X\ltimes E \xrightarrow{\overline{a}} E$$

is congruent to (4.1). Consequently, both (4.1) and (4.2) induce the same map in homology. Since (4.1) is a homotopy equivalence it induces an isomorphism in homology. Thus (4.2) also induces an isomorphism in homology. By hypothesis, $E \simeq \Omega\Sigma X \ltimes \Sigma D$ so E is simply-connected. Thus the domain and range of (4.2) are simply-connected so the fact that it induces an isomorphism in homology implies that it is a homotopy equivalence by Whitehead's Theorem. □

Recall from the setup at the beginning of the section that the restriction of $\Omega\Sigma X \ltimes \Sigma A \xrightarrow{\theta} E$ to ΣA is a map $g\colon \Sigma A \longrightarrow E$ that lifts f through p.

THEOREM 4.6. *Suppose that there is a homotopy fibration sequence $\Omega\Sigma X \xrightarrow{\partial} E \xrightarrow{p} Y \xrightarrow{h} \Sigma X$ with the following properties:*

(a) *Ωh has a right homotopy inverse;*
(b) *there is a map $\delta\colon \Sigma D \longrightarrow E$ such that the composite*

$$\Omega\Sigma X \ltimes \Sigma D \xrightarrow{1\ltimes\delta} \Omega\Sigma X \ltimes E \xrightarrow{\overline{a}} E$$

is a homotopy equivalence;

(c) *g can be chosen to factor as a composite*

$$g\colon \Sigma A \xrightarrow{\ell} X^{\wedge k} \wedge \Sigma D \xrightarrow{\overline{\mathfrak{c}}_k} \Omega\Sigma X \ltimes E \xrightarrow{\overline{a}} E$$

for some map ℓ.

Let C be the homotopy cofibre of ℓ. Then there is a homotopy equivalence

$$(\Omega\Sigma X \ltimes C) \vee (J_{k-1}(X) \ltimes \Sigma D) \longrightarrow E'.$$

PROOF. Consider the diagram

$$\begin{array}{ccccccc}
\Omega\Sigma X \ltimes \Sigma A & = & = & = & \Omega\Sigma X \ltimes \Sigma A & & \\
\Big\downarrow{\scriptstyle (1\ltimes\ell)+*} & & & & \Big\downarrow{\scriptstyle 1\ltimes g} & \searrow{\scriptstyle \theta} & \\
\Omega\Sigma X \ltimes (X^{\wedge k} \wedge \Sigma D) \vee (J_{k-1}(X) \ltimes \Sigma D) & \xrightarrow{(1\ltimes\overline{\mathfrak{c}}_k)\perp\overline{J}_k} & \Omega\Sigma X \ltimes (\Omega\Sigma X \ltimes E) & \xrightarrow{1\ltimes\overline{a}} & \Omega\Sigma X \ltimes E & \xrightarrow{\overline{a}} & E.
\end{array}$$

Since $\Omega\Sigma X \ltimes \Sigma A$ maps trivially to $J_{k-1}(X) \ltimes \Sigma D$, to show the rectangle homotopy commutes it suffices to show that $(1\ltimes\overline{a})\circ(1\ltimes\overline{\mathfrak{c}}_k)\circ(1\ltimes\ell) \simeq 1\ltimes g$. But this holds by hypothesis (c). The right triangle homotopy commutes by (2.6). As hypotheses (a) and (b) hold, Proposition 4.5 implies that the composite along the bottom row is a homotopy equivalence. Rewriting, there is a homotopy commutative square

$$\begin{array}{ccc}
\Omega\Sigma X \ltimes \Sigma A & = = = & \Omega\Sigma X \ltimes \Sigma A \\
\Big\downarrow{\scriptstyle (1\ltimes\ell)+*} & & \Big\downarrow{\scriptstyle \theta} \\
\Omega\Sigma X \ltimes (X^{\wedge k} \wedge \Sigma D) \vee (J_{k-1}(X) \ltimes \Sigma D) & \xrightarrow{\simeq} & E.
\end{array} \tag{4.3}$$

As the homotopy cofibre of ℓ is C, the homotopy cofibre of $(1 \ltimes \ell) + *$ in (4.3) is

$$(\Omega\Sigma X \ltimes C) \vee (J_{k-1}(X) \ltimes \Sigma D).$$

As the homotopy cofibre of θ is E', the homotopy commutativity of (4.3) implies that there is an induced map of cofibres

$$\alpha\colon (\Omega\Sigma X \ltimes C) \vee (J_{k-1}(X) \ltimes \Sigma D) \longrightarrow E'.$$

The homotopy equivalence in (4.3) and the five-lemma then imply that α induces an isomorphism in homology and so is a homotopy equivalence by Whitehead's Theorem. $\square$

REMARK 4.7. Note from the proof of Theorem 4.6 that the restriction of the homotopy equivalence $(\Omega\Sigma X \ltimes C) \vee (J_{k-1}(X) \ltimes \Sigma D) \longrightarrow E'$ to $J_{k-1}(X) \ltimes \Sigma D$ is homotopic to the composite

$$J_{k-1}(X) \ltimes \Sigma D \xrightarrow{\overline{J}_k} \Omega\Sigma X \ltimes (\Omega\Sigma X \ltimes E) \xrightarrow{1\ltimes\overline{a}} \Omega\Sigma X \ltimes E \xrightarrow{\overline{a}} E \longrightarrow E'. \tag{4.4}$$

It will be useful for later to give an alternative description of this composite. Consider the diagram

$$\begin{array}{ccccccc}
J_{k-1}(X) \ltimes \Sigma D & \xrightarrow{j_{k-1}\ltimes\delta} & \Omega\Sigma X \ltimes E & & \xrightarrow{\overline{a}} & E & \\
\Big\| & & \Big\downarrow{\scriptstyle j} & & & \Big\downarrow{\scriptstyle j} & \searrow\!\!\searrow \\
J_{k-1}(X) \ltimes \Sigma D & \xrightarrow{\overline{J}_k} & \Omega\Sigma X \ltimes (\Omega\Sigma X \ltimes E) & \xrightarrow{1\ltimes\overline{a}} & & \Omega\Sigma X \ltimes E & \xrightarrow{\overline{a}} E.
\end{array}$$

The left square commutes by definition of $\overline{J}_k$, the middle square commutes by the naturality of j, and the right square homotopy commutes since $\overline{a}$ is a quotient of the action a and a restricts to the identity map on E. The diagram therefore implies that (4.4) is homotopic to the composite $\overline{a}\circ(j_{k-1}\ltimes\delta)$. Writing $j_{k-1}\ltimes\delta$ as $(1\ltimes\delta)\circ(j_{k-1}\ltimes 1)$, we conclude that (4.4) is homotopic to the composite

$$J_{k-1}(X) \ltimes \Sigma D \xrightarrow{j_{k-1}\ltimes 1} \Omega\Sigma X \ltimes \Sigma D \xrightarrow{\overline{a}\circ(1\ltimes\delta)} E \longrightarrow E'$$

where $\overline{a}\circ(1\ltimes\delta)$ is assumed to be a homotopy equivalence in hypothesis (c) of Theorem 4.6.

An interesting general example of Theorem 4.6 is the following. Start with the homotopy fibration sequence

$$\Omega\Sigma X \longrightarrow E \xrightarrow{p} \Sigma X \vee \Sigma Y \xrightarrow{q_1} \Sigma X.$$

Fix a positive integer k and take $A = X^{\wedge k} \wedge Y$. Recall the definition of M_k as the homotopy cofibration

$$X^{\wedge k} \wedge \Sigma Y \xrightarrow{ad^k(i_1)(i_2)} \Sigma X \vee \Sigma Y \longrightarrow M_k.$$

Observe that q_1 extends to a map $q'_k \colon M_k \longrightarrow \Sigma X$. By Example 3.6 there is a map $\Sigma Y \xrightarrow{\delta} E$ lifting i_2 through p and for which there is a homotopy equivalence

$$\Omega\Sigma X \ltimes \Sigma Y \xrightarrow{1\ltimes\delta} \Omega\Sigma X \ltimes E \xrightarrow{\overline{a}} E.$$

(The map $\Sigma Y \longrightarrow E$ was called "g" in Example 3.6 for use in Theorem 3.5 but in the setup for Theorem 4.6 we are about to consider this map will play the role of δ and the Whitehead product $ad^{k-1}(i_1)(i_2)$ will play the role of g.) Let p'_k be the composite

$$p'_k \colon J_{k-1}(X) \ltimes \Sigma Y \xrightarrow{j_{k-1}\ltimes 1} \Omega\Sigma X \ltimes \Sigma Y \xrightarrow{\overline{a}\circ(1\ltimes\delta)} E \xrightarrow{p} \Sigma X \vee \Sigma Y \longrightarrow M_k.$$

LEMMA 4.8. *For $k \geq 1$, there is a homotopy fibration*

$$J_{k-1}(X) \ltimes \Sigma Y \xrightarrow{p'_k} M_k \xrightarrow{q'_k} \Sigma X$$

which splits after looping to give a homotopy equivalence

$$\Omega M_k \simeq \Omega\Sigma X \times \Omega(J_{k-1}(X) \ltimes \Sigma Y).$$

PROOF. We wish to apply Theorem 4.6 to the homotopy fibration sequence

$$\Omega\Sigma X \longrightarrow E \xrightarrow{p} \Sigma X \vee \Sigma Y \xrightarrow{q_1} \Sigma X$$

with appropriate choices of the maps δ and g. To do so, hypotheses (a) to (c) for the theorem need to be checked. As q_1 has a right homotopy inverse so does Ωq_1, and therefore hypothesis (a) holds. The map δ has the property that the composite $\Omega\Sigma X \ltimes \Sigma Y \xrightarrow{1\ltimes\delta} \Omega\Sigma X \ltimes E \xrightarrow{\overline{a}} E$ is a homotopy equivalence so hypothesis (b) holds. Notice that with this map δ the definition of $\overline{\mathfrak{d}}_k$ identifies with the definition of $\mathfrak{d}_k$ in Chapter 3. Thus, relabelling $\mathfrak{d}_k$ in Lemma 3.14 by $\overline{\mathfrak{d}}_k$, and using the definition of $\overline{\mathfrak{d}}_k$ as $\overline{a} \circ \overline{\mathfrak{c}}_k$, by Lemma 3.14 there is a homotopy commutative diagram

$$\begin{array}{ccccc} X^{\wedge k} \wedge \Sigma Y & \xrightarrow{\overline{\mathfrak{c}}_k} & \Omega\Sigma X \ltimes E & \xrightarrow{\overline{a}} & E \\ & \searrow^{ad^k(i_1)(i_2)} & & & \downarrow p \\ & & & & \Sigma X \vee \Sigma Y. \end{array}$$

Thus a map g lifting $ad^k(i_1)(i_2)$ through p is $\overline{a} \circ \overline{\mathfrak{c}}_k$. Taking ℓ to be the identity map on $X^{\wedge k} \wedge \Sigma Y$, the factorization of g as $\overline{a} \circ \overline{\mathfrak{c}}_k \circ \ell$ satisfies hypothesis (c). Therefore all the hypotheses of Theorem 4.6 hold. Since ℓ is the identity map on $X^{\wedge k} \wedge \Sigma Y$, its homotopy cofibre C is a point. Thus, by Theorem 4.6, if E' is the homotopy fibre of $M_k \xrightarrow{q'_k} \Sigma X$, then there is a homotopy equivalence $J_{k-1}(X) \ltimes \Sigma Y \simeq E'$. The fact that Ωq_1 has a right homotopy inverse then implies that there are homotopy equivalences

$$\Omega M \simeq \Omega\Sigma X \times \Omega E' \simeq \Omega\Sigma X \times \Omega(J_{k-1}(X) \ltimes \Sigma Y).$$

It remains to identify the composite $J_{k-1}(X) \ltimes \Sigma Y \xrightarrow{\simeq} E' \longrightarrow M_k$ as p'_k. By Remark 4.7, the homotopy equivalence $J_{k-1}(X) \ltimes \Sigma Y \simeq E'$ is realized by the composite

$$J_{k-1}(X) \ltimes \Sigma Y \xrightarrow{j_{k-1} \ltimes 1} \Omega\Sigma X \ltimes \Sigma Y \xrightarrow{\overline{a} \circ (1 \ltimes \delta)} E \longrightarrow E'.$$

Composing with $E' \longrightarrow M_k$ and using the fact that $E \longrightarrow E' \longrightarrow M_k$ is homotopic to $E \xrightarrow{p} \Sigma X \vee \Sigma Y \longrightarrow M_k$ then shows that $J_{k-1}(X) \ltimes \Sigma Y \xrightarrow{\simeq} E' \longrightarrow M_k$ is homotopic to the composite

$$J_{k-1}(X) \ltimes \Sigma Y \xrightarrow{j_{k-1} \ltimes 1} \Omega\Sigma X \ltimes \Sigma Y \xrightarrow{\overline{a} \circ (1 \ltimes \delta)} E \xrightarrow{\quad p \quad} \Sigma X \vee \Sigma Y \longrightarrow M_k$$

which is the definition of p'_k. □

In particular, observe that if $k = 1$ then we have attached $ad(i_1)(i_2) = [i_1, i_2]$ so $M_1 = \Sigma X \times \Sigma Y$, and $\bigvee_{k=0}^{0} X^k \wedge \Sigma Y = \Sigma Y$, so we recover the usual homotopy fibration $\Sigma Y \longrightarrow \Sigma X \times \Sigma Y \longrightarrow \Sigma X$.

Lemma 4.8 identifies the homotopy fibre of q'_k as $J_{k-1}(X) \ltimes \Sigma Y$ and the map from the fibre to the total space as p'_k, but we would like an alternate description of the fibre that identifies the map from it to M_k as a wedge sum of Whitehead products. Since p'_k depends on the inclusion $J_{k-1}(X) \xrightarrow{j_{k-1}} \Omega\Sigma X$ it blends well with the multiplication on $\Omega\Sigma X$, so we will make use of congruence again to make the conversion from multiplication to Whitehead products.

Define the map γ_k by the composite

$$\gamma_k \colon \bigvee_{t=0}^{k-1} X^{\wedge t} \wedge \Sigma Y \xrightarrow{\bigvee_{t=0}^{k-1} ad^t(i_1)(i_2)} \Sigma X \vee \Sigma Y \longrightarrow M_k.$$

We now prove Theorem 1.2, restated verbatim.

THEOREM 4.9. *For $k \geq 1$, there is a homotopy fibration*

$$\bigvee_{t=0}^{k-1} X^{\wedge t} \wedge \Sigma Y \xrightarrow{\gamma_k} M_k \xrightarrow{q'_k} \Sigma X$$

which splits after looping to give a homotopy equivalence

$$\Omega M_k \simeq \Omega\Sigma X \times \Omega(\bigvee_{t=0}^{k-1} X^{\wedge k} \wedge \Sigma Y).$$

PROOF. By Lemma 3.10 the map $\bigvee_{t=0}^{\infty} X^{\wedge t} \wedge \Sigma Y \xrightarrow{c} \Omega\Sigma X \ltimes \Sigma Y$ is a homotopy equivalence. By definition, c is the wedge sum of maps c_t for $0 \leq t < \infty$, and the definition of c_t via the mulitiplication on $\Omega\Sigma X$ implies that it factors through $J_{k-1}(X) \xrightarrow{j_{k-1}} \Omega\Sigma X$ if $t \leq k-1$. Thus there is a homotopy commutative square

$$\begin{array}{ccc} \bigvee_{t=0}^{k-1} X^{\wedge t} \wedge \Sigma Y & \xrightarrow{c'_{k-1}} & J_{k-1}(X) \ltimes \Sigma Y \\ \downarrow{\scriptstyle I} & & \downarrow{\scriptstyle j_{k-1} \ltimes 1} \\ \bigvee_{t=0}^{\infty} X^{\wedge k} \wedge \Sigma Y & \xrightarrow{\quad c \quad} & \Omega\Sigma X \ltimes \Sigma Y \end{array} \tag{4.5}$$

where I is the inclusion and c'_t lifts $\bigvee_{t=0}^{k-1} c_t$ through j_{k-1}. The same argument in Lemma 3.10 that shows c is a homotopy equivalence also shows that c'_{k-1} is also a homotopy equivalence.

Next, since q'_k factors through q_1 there is a homotopy fibration diagram

$$
\text{(4.6)}\qquad
\begin{array}{ccc}
E & \xrightarrow{\lambda} & J_{k-1}(X) \ltimes \Sigma Y \\
\downarrow{\scriptstyle p} & & \downarrow{\scriptstyle p'_k} \\
\Sigma X \vee \Sigma Y & \longrightarrow & M_k \\
\downarrow{\scriptstyle q_1} & & \downarrow{\scriptstyle q'_k} \\
\Sigma X & = & \Sigma X
\end{array}
$$

for some induced map λ of fibres. Consider the composite

$$\kappa\colon J_{k-1}(X) \ltimes \Sigma Y \xrightarrow{j_{k-1}\ltimes 1} \Omega\Sigma X \ltimes \Sigma Y \xrightarrow{\overline{a}\circ(1\ltimes\delta)} E \xrightarrow{\lambda} J_{k-1}(X) \ltimes \Sigma Y.$$

We claim that κ is homotopic to the identity map. To see this, observe that by (4.6), $p'_k \circ \kappa$ is homotopic to the composite $J_{k-1}(X) \ltimes \Sigma Y \xrightarrow{j_{k-1}\ltimes 1} \Omega\Sigma X \ltimes \Sigma Y \xrightarrow{\overline{a}\circ(1\ltimes\delta)} E \xrightarrow{p} \Sigma X \vee \Sigma Y \longrightarrow M_k$, which is the definition of p'_k. Thus $p'_k \circ \kappa \simeq p'_k$. Since $J_{k-1}(X) \ltimes \Sigma Y$ is a suspension, we may subtract in order to get $p'_k \circ (1-\kappa) \simeq *$. There is a homotopy fibration $\Omega\Sigma X \xrightarrow{s} J_{k-1}(X) \ltimes \Sigma Y \xrightarrow{p'_k} M_k$ where s is null homotopic by Lemma 4.8. The null homotopy for $p'_k \circ (1-\kappa)$ implies that $1-\kappa$ lifts through s, and hence is null homotopic. Thus κ is homotopic to the identity map.

Putting (4.5) and (4.6) together gives a homotopy commutative diagram

$$
\begin{array}{ccccccccc}
\bigvee_{t=0}^{k-1} X^{\wedge t} \wedge \Sigma Y & \xrightarrow{c'_{k-1}} & J_{k-1}(X) \ltimes \Sigma Y & \xrightarrow{\overline{a}\circ(j_{k-1}\ltimes\delta)} & E & = & E & \xrightarrow{\lambda} & J_{k-1}(X) \ltimes \Sigma Y \\
\downarrow{\scriptstyle I} & & \downarrow{\scriptstyle j_{k-1}\ltimes 1} & & \| & & \downarrow{\scriptstyle p} & & \downarrow{\scriptstyle p'_k} \\
\bigvee_{t=0}^{\infty} X^{\wedge k} \wedge \Sigma Y & \xrightarrow{c} & \Omega\Sigma X \ltimes \Sigma Y & \xrightarrow{\overline{a}\circ(1\ltimes\delta)} & E & \xrightarrow{p} & \Sigma X \vee \Sigma Y & \longrightarrow & M_k.
\end{array}
$$

Note that the middle squares commute. Along the top row both c'_{k-1} and $\kappa = \lambda\circ\overline{a}\circ(j_{k-1}\ltimes\delta)$ are homotopy equivalences. Therefore, $\lambda_*\circ(\overline{a}\circ(j_{k-1}\ltimes\delta)\circ c'_{k-1})_*$ is an isomorphism. Along the bottom row the composite $\overline{a}\circ(1\ltimes\delta)\circ c$ is the definition of the map d in Chapter 3. By Lemma 3.13, d is congruent to the map $\mathfrak{d}$. In particular, in homology $d_* = \mathfrak{d}_*$. Therefore $(\mathfrak{d}\circ I)_* = (d\circ I)_*$, which by the previous diagram equals $(\overline{a}\circ(j_{k-1}\ltimes\delta)\circ c'_{k-1})_*$. Hence $\lambda_*\circ(\mathfrak{d}\circ I)_*$ is an isomorphism, implying that $\lambda\circ\mathfrak{d}\circ I$ is a homotopy equivalence by Whitehead's Theorem.

By Theorem 3.15 there is a homotopy commutative diagram

$$
\begin{array}{ccccccc}
\bigvee_{t=0}^{k-1} X^{\wedge t} \wedge \Sigma Y & \xrightarrow{I} & \bigvee_{t=0}^{\infty} X^{\wedge t} \wedge \Sigma Y & \xrightarrow{\mathfrak{d}} & E & \xrightarrow{\lambda} & J_{k-1}(X) \ltimes \Sigma Y \\
& & & \searrow{\scriptstyle \bigvee_{t=0}^{\infty} ad^t(i_1)(i_2)} & \downarrow{\scriptstyle p} & & \downarrow{\scriptstyle p'_k} \\
& & & & \Sigma X \vee \Sigma Y & \longrightarrow & M_k.
\end{array}
$$

As the top row of this diagram is a homotopy equivalence and the bottom direction around the diagram matches the definition of γ_k, the homotopy fibration

$$J_{k-1}(X) \ltimes \Sigma Y \xrightarrow{p'_k} M_k \xrightarrow{q'_k} \Sigma X$$

may be replaced up to equivalence by a homotopy fibration $\bigvee_{t=0}^{k-1} X^{\wedge t} \wedge \Sigma Y \xrightarrow{\gamma_k} M_k \xrightarrow{q'_k} \Sigma X$. This proves the first assertion of the lemma, the splitting of the fibration after looping follows from the existence of a right homotopy inverse of $\Omega q'_k$. □

Interesting specific examples occur when $X = S^m$ and $Y = S^n$. Note then that for each $k \geq 1$ we have $X^{\wedge k} \wedge \Sigma Y \simeq S^{km+n+1}$.

EXAMPLE 4.10. For $k \geq 1$, define M_k by the homotopy cofibration

$$S^{km+n+1} \xrightarrow{ad^k(i_1)(i_2)} S^{m+1} \vee S^{n+1} \longrightarrow M_k.$$

Then there is a homotopy fibration

$$\bigvee_{t=0}^{k-1} S^{tm+n+1} \xrightarrow{\gamma_k} M_k \longrightarrow S^{m+1}$$

where γ_k is the composite

$$\bigvee_{t=0}^{k-1} S^{tm+n+1} \xrightarrow{\bigvee_{t=0}^{k-1} ad^t(i_1)(i_2)} S^{m+1} \vee S^{n+1} \longrightarrow M_k,$$

and after looping this homotopy fibration splits to give a homotopy equivalence

$$\Omega M_k \simeq \Omega S^{m+1} \times \Omega\Big(\bigvee_{t=0}^{k-1} S^{tm+n+1}\Big).$$

In particular, if $k = 2$ then M_2 is the homotopy cofibre of $[i_1, [i_1, i_2]]$ and there is a homotopy equivalence $\Omega M_2 \simeq \Omega S^{m+1} \times \Omega(S^{n+1} \vee S^{m+n+1})$.

A bit more generally, take $X = S^m$ and let Y be arbitrary. Note then that for each $k \geq 1$ we have $X^{\wedge k} \wedge \Sigma Y \simeq \Sigma^{km+1} Y$.

EXAMPLE 4.11. For $k \geq 1$, define M_k by the homotopy cofibration

$$\Sigma^{km+1} Y \xrightarrow{ad^k(i_1)(i_2)} S^{m+1} \vee \Sigma Y \longrightarrow M_k.$$

Then there is a homotopy fibration

$$\bigvee_{t=0}^{k-1} \Sigma^{tm+1} Y \xrightarrow{\gamma_k} M_k \longrightarrow S^{m+1}$$

where γ_k is the composite

$$\bigvee_{t=0}^{k-1} \Sigma^{tm+1} Y \xrightarrow{\bigvee_{t=0}^{k-1} ad^t(i_1)(i_2)} S^{m+1} \vee \Sigma Y \longrightarrow M_k,$$

and after looping this homotopy fibration splits to give a homotopy equivalence

$$\Omega M_k \simeq \Omega S^{m+1} \times \Omega\Big(\bigvee_{t=0}^{k-1} \Sigma^{tm+1} Y\Big).$$

In particular, if $k = 2$ then M_2 is the homotopy cofibre of $[i_1, [i_1, i_2]]$ and there is a homotopy equivalence $\Omega M_2 \simeq \Omega S^{m+1} \times \Omega(\Sigma Y \vee \Sigma^{m+1} Y)$.

Even more can be said about the homotopy theory of the spaces M_k. Return to the general case, of a homotopy cofibration $X^{\wedge k} \wedge \Sigma Y \xrightarrow{ad^k(i_1)(i_2)} \Sigma X \vee \Sigma Y \xrightarrow{m_k} M_k$, where the inclusion of $\Sigma X \vee \Sigma Y$ into M_k is now labelled m_k. In Corollary 4.13 we identify the homotopy fibre of m_k. To compress notation, write ad^k for $ad^k(i_1)(i_2)$.

LEMMA 4.12. *The homotopy cofibration* $X^{\wedge k} \wedge \Sigma Y \xrightarrow{ad^k} \Sigma X \vee \Sigma Y \xrightarrow{m_k} M_k$ *has the property that the map* Ωm_k *has a right homotopy inverse.*

PROOF. By Theorem 4.9, there is a homotopy fibration

$$\bigvee_{t=0}^{k-1} X^{\wedge t} \wedge \Sigma Y \xrightarrow{\gamma_k} M_k \xrightarrow{q'_k} \Sigma X$$

which splits after looping to give a homotopy equivalence

$$\Omega M_k \simeq \Omega\Sigma X \times \Omega(\bigvee_{t=0}^{k-1} X^{\wedge k} \wedge \Sigma Y).$$

Here, γ_k is the composite

$$\bigvee_{t=0}^{k-1} X^{\wedge t} \wedge \Sigma Y \xrightarrow{\bigvee_{t=0}^{k-1} ad^t} \Sigma X \vee \Sigma Y \xrightarrow{m_k} M_k,$$

a right homotopy inverse for q'_k is the composite

$$i'_1 \colon \Sigma X \xrightarrow{i_1} \Sigma X \vee \Sigma Y \xrightarrow{m_k} M_k,$$

and the homotopy equivalence is given by the composite

$$\Omega\Sigma X \times \Omega(\bigvee_{i=0}^{k-1} X^{\wedge k} \wedge \Sigma Y) \xrightarrow{\Omega i'_1 \times \Omega\gamma_k} \Omega M_k \times \Omega M_k \xrightarrow{\mu} \Omega M_k \tag{4.7}$$

where μ is the standard loop multiplication. Observe that both i'_1 and γ_k factor through $\Sigma X \vee \Sigma Y$, so as Ωm_k is an H-map the homotopy equivalence in (4.7) is homotopic to the composite

$$\Omega\Sigma X \times \Omega(\bigvee_{i=0}^{k-1} X^{\wedge k} \wedge \Sigma Y) \xrightarrow{\Omega i_1 \times \Omega a_k} \Omega(\Sigma X \vee \Sigma Y) \times \Omega(\Sigma X \vee \Sigma Y) \xrightarrow{\mu} \Omega(\Sigma X \vee \Sigma Y) \xrightarrow{\Omega m_k} \Omega M_k$$

where $a_k = \bigvee_{i=0}^{k-1} ad^i(i_1)(i_2)$. In particular, this implies that Ωm_k has a right homotopy inverse. □

By Lemma 4.12, Ωm_k has a right homotopy inverse $s \colon \Omega M_k \longrightarrow \Omega(\Sigma X \vee \Sigma Y)$. The existence of s implies that the hypotheses of Theorem 3.5 are satisfied when applied to the homotopy cofibration $X^{\wedge k} \wedge \Sigma Y \xrightarrow{ad^k} \Sigma X \vee \Sigma Y \xrightarrow{m_k} M_k$. Therefore we immediately obtain the following.

COROLLARY 4.13. *There is a homotopy fibration*

$$\Omega M_k \ltimes (X^{\wedge k} \wedge \Sigma Y) \xrightarrow{\chi} \Sigma X \vee \Sigma Y \xrightarrow{m_k} M_k$$

where χ *is the sum of the maps*

$$\Omega M_k \ltimes (X^{\wedge k} \wedge \Sigma Y) \xrightarrow{\pi} X^{\wedge k} \wedge \Sigma Y \xrightarrow{ad^k} \Sigma X \vee \Sigma Y$$

and

$$\Omega M_k \ltimes (X^{\wedge k} \wedge \Sigma Y) \xrightarrow{q} \Omega M_k \wedge (X^{\wedge k} \wedge \Sigma Y) \xrightarrow{[ev \circ s, ad^k]} \Sigma X \vee \Sigma Y.$$ □

EXAMPLE 4.14. Consider the homotopy cofibration

$$S^{km+n+1} \xrightarrow{ad^k(i_1)(i_2)} S^{m+1} \vee S^{n+1} \xrightarrow{m_k} M_k$$

from Example 4.10. By Lemma 4.12 the map Ωm_k has a right homotopy inverse and by Theorem 3.5 there is a homotopy fibration

$$\Omega M_k \ltimes S^{km+n+1} \xrightarrow{\chi} S^{m+1} \vee S^{n+1} \xrightarrow{m_k} M_k$$

where χ is the sum of $\Omega M_k \ltimes S^{km+n+1} \xrightarrow{\pi} S^{km+n+1} \xrightarrow{ad^k} S^{m+1} \vee S^{n+1}$ and $\Omega M_k \ltimes S^{km+n+1} \longrightarrow \Omega M_k \wedge S^{km+n+1} \xrightarrow{[ev\circ s, ad^k]} S^{m+1} \vee S^{n+1}$.

Relation to Moore's Conjecture. A few definitions are necessary to state the conjecture.

DEFINITION 4.15. Let X be a simply-connected CW-complex and let p be a prime. The *homotopy exponent* of X at p is the least power of p that annihilates the p-torsion in $\pi_*(X)$.

Write $\exp_p(X) = p^r$ if p^r is this least power of p. If the prime is understood this may be shortened to $\exp(X) = p^r$. If $\pi_*(X)$ has torsion of all orders, write $\exp_p(X) = \infty$.

DEFINITION 4.16. Let X be a simply-connected CW-complex. If there are finitely many $\mathbb{Z}$ summands in $\pi_*(X)$ then X is *elliptic*, otherwise X is *hyperbolic*.

CONJECTURE 4.17 (Moore). Let X be a simply-connected finite CW-complex. Then the following are equivalent:

- X is elliptic;
- $\exp_p(X)$ is finite for some prime p;
- $\exp_p(X)$ is finite for all primes p.

Moore's Conjecture posits a remarkable relationship between the rational homotopy groups of X and its torsion homotopy groups. The rational homotopy groups deeply influence the torsion homotopy groups, and torsion at any one prime deeply influences the torsion that occurs at any prime. The conjecture has been shown to hold in a wide variety of cases: for H-spaces [**L**], for torsion-free suspensions [**Se**], for various 2 and 3-cell complexes [**NS**], for generalized moment-angle complexes [**HST**], and for families of highly-connected Poincaré duality complexes [**Ba, BB, BT1, BT2**].

More examples of spaces for which Moore's Conjecture holds may be extracted from Proposition 4.9. We give three examples.

EXAMPLE 4.18. Return to Example 4.10. The $k=1$ case is $M_k \simeq S^{m+1} \times S^{n+1}$. A sphere was shown to have a finite homotopy exponent for any prime p by James [**J2**] for $p = 2$ and Toda [**To**] for any odd prime. Therefore the product of a finite number of spheres also has an exponent for any prime p, and of course, this product is elliptic. The case when the attaching map $[i_1, i_2]$ for the top cell in $S^{m+1} \times S^{n+1}$ is replaced by $q \cdot [i_1, i_2]$ for a prime q a nonzero integer was considered by Neisendorfer and Selick [**NS**] and shown to also be elliptic and have an exponent at every prime p. Now suppose that $k \geq 2$. An argument using the Hilton-Milnor Theorem shows that a wedge of spheres is hyperbolic and has no exponent at any prime (see, for example, [**NS**]). Example 4.10 shows that $\Omega M_k \simeq \Omega S^{m+1} \times \Omega(\bigvee_{t=0}^{k-1} S^{tm+n+1})$. In particular, $\Omega(S^{n+1} \vee S^{m+n+1})$ is a retract of ΩM_k, and so M_k must be hyperbolic and have no exponent at any prime.

REMARK 4.19. Using different methods, Anick [**A**, Lemma 2.3 and Theorem 3.7] showed that any space obtained by attaching a sphere to a wedge of two or more spheres by a linear combination of Whitehead products satisfies Moore's Conjecture for all primes except possibly 2 and 3. In particular, this is true of the spaces M_k for $k \geq 2$. Example 4.18 is an improvement in this case as there is no restriction on the primes. Further, Example 4.10 goes much further by giving an explicit integral homotopy decomposition of ΩM_k.

Recall from the Introduction that for $n \geq 2$, p a prime and $r \geq 1$, the *mod-p^r Moore space* $P^n(p^r)$ is defined as the homotopy cofibre of the degree p^r map on S^{n-1}. Note that $\Sigma P^n(p^r) \simeq P^{n+1}(p^r)$. A useful property that will be needed at several points is a result of Neisendorfer [**N1**, Corollary 6.6] that describes the homotopy type of the smash product of two mod-p^r Moore spaces.

LEMMA 4.20. *Let p be a prime, r a nonnegative integer, and assume that $p^r \neq 2$. If $s, t \geq 2$ then there is a homotopy equivalence*

$$P^s(p^r) \wedge P^t(p^r) \simeq P^{s+t}(p^r) \vee P^{s+t-1}(p^r). \qquad \square$$

The $p^r = 2$ case is very different: the smash product $P^s(2) \wedge P^t(2)$ is known to be indecomposable.

By [**N3**], if $n \geq 3$ and p is odd then $\exp_p(P^n(p^r)) = p^{r+1}$; by [**Th**], if $n \geq 4$, $p = 2$ and $r \geq 6$ then $\exp_2(P^n(2^r)) = 2^{r+1}$, and by [**C**], if $n \geq 3$, $p = 2$ and $r \geq 2$ then $P^n(2^r)$ has a finite 2-primary exponent. In all cases, as $P^n(p^r)$ is contractible when localized at a prime $q \neq p$, or rationally, we see that $P^n(p^r)$ is elliptic and has a finite homotopy exponent at every prime p and so satisfies Moore's Conjecture.

EXAMPLE 4.21. Return to Example 4.11. Take $Y = P^n(p^r)$ for $n \geq 3$ and $p^r \neq 2$. The example shows that

$$\Omega M_k \simeq \Omega S^{m+1} \times \Omega(\bigvee_{t=0}^{k-1} P^{tm+n+1}(p^r)).$$

Using the Hilton-Milnor Theorem and Lemma 4.20 iteratively shows that the loops on a wedge of mod-p^r Moore spaces with $p^r \neq 2$ is homotopy equivalent to a finite type infinite product of mod-p^r Moore spaces. Consequently, M_k is elliptic and has a finite exponent at every prime p. Hence M_k satisfies Moore's Conjecture.

EXAMPLE 4.22. In Theorem 4.9 take $X = P^m(p^r)$ and $Y = P^n(p^r)$ for $m, n \geq 3$ and $p^r \neq 2$. Then by Lemma 4.20 there is a homotopy equivalence

$$\Omega M_k \simeq \Omega P^{m+1}(p^r) \times \Omega W$$

where W is a finite type wedge of mod-p^r Moore spaces. Arguing as in the previous example then shows that M_k is elliptic (it is rationally trivial) and has a finite exponent at every prime p. Hence it satisfies Moore's Conjecture.

CHAPTER 5

An improvement

The factorization of $\Sigma A \xrightarrow{g} E$ through $X^{\wedge k} \wedge \Sigma D$ in Theorem 4.6 gives a condition which lets us find the homotopy type of E', but it does not apply to many of the cases we are interested in. The construction behind that theorem needs as input Whitehead products of the form $[i_1, f]$ for some map f, but if we try to attach S^{2n+1} to $S^{n+1} \vee S^{n+1} \vee S^{n+1}$ (with $X = S^n$ and $D = Y = S^n \vee S^n$) by $[i_1, i_2] + [i_2, i_3]$ then the map does not have the right form so the theorem does not apply. To handle the latter case we have to allow the attaching map to have some component in ΣD as well as $X^{\wedge k} \wedge \Sigma D$, in other words, we need to consider the case where g factors through $X^{\wedge k} \ltimes \Sigma D$.

This will require some modifications to the strategy behind proving Theorem 4.6. There were two key steps: first, take the half-smash of $\Omega\Sigma X$ with the factorization of g as $\Sigma A \xrightarrow{\ell} X^{\wedge k} \wedge \Sigma D \xrightarrow{\overline{\mathfrak{c}}_k} \Omega\Sigma X \ltimes \Sigma D \xrightarrow{\overline{a}} E$ to obtain a homotopy commutative diagram

$$\begin{array}{ccccccc}
\Omega\Sigma X \ltimes \Sigma A & = & = & = & \Omega\Sigma X \ltimes \Sigma A & & \\
\big\downarrow{\scriptstyle (1\ltimes \ell)} & & & & \big\downarrow{\scriptstyle 1\ltimes g} & \searrow^{\theta} & \\
\Omega\Sigma X \ltimes (X^{\wedge k} \wedge \Sigma D)) & \xrightarrow{1\ltimes \overline{\mathfrak{c}}_k} & \Omega\Sigma X \ltimes (\Omega\Sigma X \ltimes E) & \xrightarrow{1\ltimes \overline{a}} & \Omega\Sigma X \ltimes E & \xrightarrow{\overline{a}} & E.
\end{array}$$

Second, adjust the bottom row by inserting the wedge $J_{k-1} \ltimes \Sigma D$ via the map $\overline{J}_k$ in order to obtain a homotopy equivalence along the bottom row.

To modify this, first take the half-smash of $\Omega\Sigma X$ with a factorization of g as $\Sigma A \xrightarrow{\ell} X^{\wedge k} \ltimes \Sigma D \xrightarrow{\overline{\mathfrak{c}}'_k} \Omega\Sigma X \ltimes \Sigma D \xrightarrow{\overline{a}} E$ for an appropriate map $\overline{\mathfrak{c}}'_k$ to obtain a homotopy commutative diagram

$$\begin{array}{ccccccc}
\Omega\Sigma X \ltimes \Sigma A & = & = & = & \Omega\Sigma X \ltimes \Sigma A & & \\
\big\downarrow{\scriptstyle (1\ltimes \ell)} & & & & \big\downarrow{\scriptstyle 1\ltimes g} & \searrow^{\theta} & \\
\Omega\Sigma X \ltimes (X^{\wedge k} \ltimes \Sigma D)) & \xrightarrow{1\ltimes \overline{\mathfrak{c}}'_k} & \Omega\Sigma X \ltimes (\Omega\Sigma X \ltimes E) & \xrightarrow{1\ltimes \overline{a}} & \Omega\Sigma X \ltimes E & \xrightarrow{\overline{a}} & E.
\end{array}$$

Then the bottom row has to be adjusted to obtain a homotopy equivalence. This adjustment involves more than just inserting an extra space, it also involves removing part of $\Omega\Sigma X \ltimes (X^{\wedge k} \ltimes \Sigma D)$, and this requires some extra hypotheses. The precise statement generalizing Theorem 4.6 is Theorem 5.8.

In general, let B, C and D be path-connected, pointed spaces. Define e_1, e_2 and e_3 by the composites

$$e_1\colon B \ltimes (C \wedge \Sigma D) \xrightarrow{1\ltimes i} B \ltimes (C \ltimes \Sigma D)$$

$$e_2\colon \Sigma D \xrightarrow{j} C \ltimes \Sigma D \xrightarrow{j} B \ltimes (C \ltimes \Sigma D)$$

$$e_3\colon B \wedge \Sigma D \xrightarrow{i} B \ltimes \Sigma D \xrightarrow{1\ltimes j} B \ltimes (C \ltimes \Sigma D).$$

Define maps f_1, f_2 and f_3 by the composites

$$f_1\colon B \ltimes (C \ltimes \Sigma D) \xrightarrow{1\ltimes q} B \ltimes (C \wedge \Sigma D)$$

$$f_2\colon B \ltimes (C \ltimes \Sigma D) \xrightarrow{\pi} C \ltimes \Sigma D \xrightarrow{\pi} \Sigma D$$

$$f_3\colon B \ltimes (C \ltimes \Sigma D) \xrightarrow{1\ltimes\pi} B \ltimes \Sigma D \xrightarrow{q} B \wedge \Sigma D.$$

LEMMA 5.1. *For $1 \leq i, j \leq 3$, the composite $f_i \circ e_i$ is homotopic to the identity map while if $i \neq j$ then $f_j \circ e_i$ is null homotopic.*

PROOF. In general, the composites $\Sigma D \xrightarrow{j} B \ltimes \Sigma D \xrightarrow{\pi} \Sigma D$ and $B \wedge \Sigma D \xrightarrow{i} B \ltimes \Sigma D \xrightarrow{q} B \wedge \Sigma D$ are homotopic to the identity maps while the composites $\Sigma D \xrightarrow{j} B \ltimes \Sigma D \xrightarrow{q} B \wedge \Sigma D$ and $B \wedge \Sigma D \xrightarrow{i} B \ltimes \Sigma D \xrightarrow{\pi} \Sigma D$ are null homotopic. The assertions now follow from the definitions of maps e_i and f_i for $1 \leq i \leq 3$. □

The wedge sum of e_1, e_2 and e_3 is a map

$$e\colon (B \ltimes (C \wedge \Sigma D)) \vee \Sigma D \vee (B \wedge \Sigma D) \longrightarrow B \ltimes (C \ltimes \Sigma D).$$

LEMMA 5.2. *The map e is a homotopy equivalence.*

PROOF. The wedge sum of $C \wedge \Sigma D \xrightarrow{i} C \ltimes \Sigma D$ and $\Sigma D \xrightarrow{j} C \ltimes \Sigma D$ is a homotopy equivalence. Therefore, taking half-smashes with B, the wedge sum of $B \ltimes (C \wedge \Sigma D) \xrightarrow{1\ltimes i} B \ltimes (C \ltimes \Sigma D)$ and $B \ltimes \Sigma D \xrightarrow{1\ltimes j} B \ltimes (C \ltimes \Sigma D)$ is a homotopy equivalence. Notice that $1 \ltimes i$ is the definition of e_1. Next, consider $1 \ltimes j$. The wedge sum of $\Sigma D \xrightarrow{j} B \ltimes \Sigma D$ and $B \wedge \Sigma D \xrightarrow{i} B \ltimes \Sigma D$ is a homotopy equivalence. So $1 \ltimes j$ may be rewritten as the wedge sum of the composites $\Sigma D \xrightarrow{j} B \ltimes \Sigma D \xrightarrow{j} B \ltimes (C \ltimes \Sigma D)$ and $B \wedge \Sigma D \xrightarrow{i} B \ltimes \Sigma D \xrightarrow{1\ltimes j} B \ltimes (C \ltimes \Sigma D)$, that is, $1 \ltimes j$ may be rewritten as the wedge sum of e_2 and e_3. Therefore the wedge sum of e_1, e_2 and e_3 (that is, e) is a homotopy equivalence. □

In our case, we start as in Theorem 4.6 with a homotopy fibration sequence $\Omega\Sigma X \xrightarrow{\partial} E \xrightarrow{p} Y \xrightarrow{h} \Sigma X$ such that Ωh has a right homotopy inverse and there is a map $\delta\colon \Sigma D \longrightarrow E$ such that the composite

$$\Omega\Sigma X \ltimes \Sigma D \xrightarrow{1\ltimes\delta} \Omega\Sigma X \ltimes E \xrightarrow{\overline{a}} E$$

is a homotopy equivalence. Assume there is a map $\Sigma A \longrightarrow Y$ that lifts through p to $\Sigma A \xrightarrow{g} Y$.

The construction of the maps e_i and f_i above in this case are composites

$$e_1\colon \Omega\Sigma X \ltimes (X^{\wedge k} \wedge \Sigma D) \xrightarrow{1\ltimes i} \Omega\Sigma X \ltimes (X^{\wedge k} \ltimes \Sigma D)$$

$$e_2\colon \Sigma D \xrightarrow{j} X^{\wedge k} \ltimes \Sigma D \xrightarrow{j} \Omega\Sigma X \ltimes (X^{\wedge k} \ltimes \Sigma D)$$

$$e_3\colon \Omega\Sigma X \wedge \Sigma D \xrightarrow{i} \Omega\Sigma X \ltimes \Sigma D \xrightarrow{1\ltimes j} \Omega\Sigma X \ltimes (X^{\wedge k} \ltimes \Sigma D)$$

and

$$f_1\colon \Omega\Sigma X \ltimes (X^{\wedge k} \ltimes \Sigma D) \xrightarrow{1\ltimes q} \Omega\Sigma X \ltimes (X^{\wedge k} \wedge \Sigma D)$$

$$f_2\colon \Omega\Sigma X \ltimes (X^{\wedge k} \ltimes \Sigma D) \xrightarrow{\pi} X^{\wedge k} \ltimes \Sigma D \xrightarrow{\pi} \Sigma D$$

$$f_3\colon \Omega\Sigma X \ltimes (X^{\wedge k} \ltimes \Sigma D) \xrightarrow{1\ltimes\pi} \Omega\Sigma X \ltimes \Sigma D \xrightarrow{q} \Omega\Sigma X \wedge \Sigma D.$$

By Lemma 5.1, for $1 \le i, j \le 3$ the composite $f_i \circ e_i$ is homotopic to the identity map and if $i \ne j$ the composite $f_j \circ e_i$ is null homotopic. By Lemma 5.2, the wedge sum of e_1, e_2 and e_3 gives a homotopy equivalence

$$e\colon (\Omega\Sigma X \ltimes (X^{\wedge k} \wedge \Sigma D)) \vee \Sigma D \vee (\Omega\Sigma X \wedge \Sigma D) \longrightarrow \Omega\Sigma X \ltimes (X^{\wedge k} \rtimes \Sigma D).$$

Given a map $\Sigma A \xrightarrow{\ell} X^{\wedge k} \rtimes \Sigma D$, let κ be the composite

$$\kappa\colon \Omega\Sigma X \ltimes \Sigma A \xrightarrow{1 \ltimes \ell} \Omega\Sigma X \ltimes (X^{\wedge k} \rtimes \Sigma D) \xrightarrow{e^{-1}} (\Omega\Sigma X \ltimes (X^{\wedge k} \wedge \Sigma D)) \vee \Sigma D \vee (\Omega\Sigma X \wedge \Sigma D).$$

By definition of κ there is a commutative square
(5.1)

$$\begin{array}{ccc}
\Omega\Sigma X \ltimes \Sigma A & = & \Omega\Sigma X \ltimes \Sigma A \\
\downarrow{\scriptstyle \kappa} & & \downarrow{\scriptstyle 1 \ltimes \ell} \\
(\Omega\Sigma X \ltimes (X^{\wedge k} \wedge \Sigma D)) \vee \Sigma D \vee (\Omega\Sigma X \wedge \Sigma D) & \xrightarrow{e} & \Omega\Sigma X \ltimes (X^{\wedge k} \rtimes \Sigma D).
\end{array}$$

By the Hilton-Milnor Theorem, $\kappa = \kappa_1 + \kappa_2 + \kappa_3 + W$ where κ_1, κ_2 and κ_3 are obtained by composing κ with the pinch maps p_1, p_2 and p_3 to $\Omega\Sigma X \ltimes (X^{\wedge k} \wedge \Sigma D)$, ΣD and $\Omega\Sigma X \wedge \Sigma D$ respectively, and W factors through a wedge sum of iterated Whitehead products.

We identify κ_1, κ_2 and κ_3. Since e is the wedge sum of e_1, e_2 and e_3, the fact that for $1 \le i, j \le 3$ the composite $f_i \circ e_i$ is homotopic to the identity map while if $i \ne j$ the composite $f_j \circ e_i$ is null homotopic implies that $f_i \circ e \simeq p_i$. Thus using $e \circ \kappa = 1 \ltimes \ell$ in (5.1) we obtain, for $1 \le i \le 3$,

$$\kappa_i = p_i \circ \kappa \simeq f_i \circ e \circ \kappa = f_i \circ (1 \ltimes \ell). \tag{5.2}$$

LEMMA 5.3. *Suppose that the composite* $\Sigma^2 A \xrightarrow{\Sigma\ell} \Sigma(X^{\wedge k} \rtimes \Sigma D) \xrightarrow{\Sigma\pi} \Sigma^2 D$ *is null homotopic. Then the maps* $\Sigma\kappa_2$ *and* $\Sigma\kappa_3$ *are null homotopic.*

PROOF. By (5.2), $\kappa_2 \simeq f_2 \circ (1 \ltimes \ell)$. Consider the diagram

$$\begin{array}{ccccc}
\Omega\Sigma X \ltimes \Sigma A & \xrightarrow{\pi} & \Sigma A & & \\
\downarrow{\scriptstyle 1 \ltimes \ell} & & \downarrow{\scriptstyle \ell} & & \\
\Omega\Sigma X \ltimes (X^{\wedge k} \rtimes \Sigma D) & \xrightarrow{\pi} & X^{\wedge k} \rtimes \Sigma D & \xrightarrow{\pi} & \Sigma D.
\end{array}$$

The square commutes by the naturality of π. As the bottom row is the definition of f_2, the lower direction around the diagram is $f_2 \circ (1 \ltimes \ell)$, that is, κ_2. This equals the upper direction around the diagram, which is null homotopic after suspending since $\Sigma\pi \circ \Sigma\ell$ is null homotopic by hypothesis. Thus $\Sigma\kappa_2$ is null homotopic.

By (5.2), $\kappa_3 \simeq f_3 \circ (1 \ltimes \ell)$. By definition, $f_3 = q \circ (1 \ltimes \pi)$. Thus κ_3 factors through $1 \ltimes (\pi \circ \ell)$. Suspending, $\Sigma\kappa_3$ factors through $\Sigma(1 \ltimes (\pi \circ \ell)) \simeq 1 \ltimes \Sigma(\pi \circ \ell)$. By hypothesis $\Sigma(\pi \circ \ell)$ is null homotopic, and therefore $\Sigma\kappa_3$ is null homotopic. □

COROLLARY 5.4. *Suppose that the composite* $\Sigma^2 A \xrightarrow{\Sigma\ell} \Sigma(X^{\wedge k} \rtimes \Sigma D) \xrightarrow{\Sigma\pi} \Sigma^2 D$ *is null homotopic. Then there is a homotopy commutative square*

$$\begin{array}{ccc}
\Sigma(\Omega\Sigma X \ltimes \Sigma A) & = & \Sigma(\Omega\Sigma X \ltimes \Sigma A) \\
\downarrow{\scriptstyle \Sigma\kappa_1} & & \downarrow{\scriptstyle \Sigma(1 \ltimes \ell)} \\
\Sigma(\Omega\Sigma X \ltimes (X^{\wedge k} \wedge \Sigma D)) & \xrightarrow{\Sigma e_1} & \Sigma(\Omega\Sigma X \ltimes (X^{\wedge k} \rtimes \Sigma D)).
\end{array}$$

PROOF. Following (5.1) we saw that $\kappa = \kappa_1 + \kappa_2 + \kappa_3 + W$ where W factors through a wedge sum of Whitehead products. In particular, as Whitehead products suspend trivially, ΣW is null homotopic. By Lemma 5.3, $\Sigma\kappa_2$ and $\Sigma\kappa_3$ are also null homotopic. Therefore $\Sigma\kappa \simeq \Sigma\kappa_1$ and the homotopy commutativity of the asserted diagram follows. □

Some maps need to be defined that modify the maps $\overline{\mathfrak{c}}_k$ and J_k from Chapter 4. For $k = 1$ define the map $\overline{\mathfrak{c}}'_1$ by

$$\overline{\mathfrak{c}}'_1\colon X \ltimes \Sigma D \xrightarrow{E\ltimes\delta} \Omega\Sigma X \ltimes E.$$

Note that $\overline{\mathfrak{c}}'_1 \circ i = \overline{\mathfrak{c}}_1$. For $k \geq 2$ a recursive definition is used: define the maps $\overline{\mathfrak{c}}'_k$ by the composite

$$\overline{\mathfrak{c}}'_k\colon X^{\wedge k} \ltimes \Sigma D \xrightarrow{q+\pi} (X^{\wedge k} \wedge \Sigma D) \vee \Sigma D \xrightarrow{\overline{\mathfrak{c}}_k \perp j\circ\delta} \Omega\Sigma X \ltimes E.$$

Also, define the map J'_k by the composite

$$J'_k\colon J_{k-1}(X) \ltimes \Sigma D \xrightarrow{j_{k-1}\ltimes j} \Omega\Sigma X \ltimes (X^{\wedge k} \ltimes \Sigma D).$$

Recall that $\overline{J}_k$ was defined in Chapter 4 as the composite

$$\overline{J}_k\colon J_{k-1}(X) \ltimes \Sigma D \xrightarrow{j_{k-1}\ltimes\delta} \Omega\Sigma X \ltimes E \xrightarrow{\ j\ } \Omega\Sigma X \ltimes (\Omega\Sigma X \ltimes E).$$

LEMMA 5.5. *The following hold:*

(a) *the composite* $\Sigma X^{\wedge k}\wedge\Sigma D \xrightarrow{\Sigma i} \Sigma X^{\wedge k}\ltimes\Sigma D \xrightarrow{\Sigma\overline{\mathfrak{c}}'_k} \Sigma\Omega\Sigma X\ltimes E$ *is homotopic to* $\Sigma\overline{\mathfrak{c}}_k$;

(b) *the composite*

$$\Sigma D \xrightarrow{j} X^{\wedge k} \ltimes \Sigma D \xrightarrow{\overline{\mathfrak{c}}'_k} \Omega\Sigma X \ltimes E$$

is homotopic to $j \circ \delta$;

(c) *the composite*

$$J_{k-1}(X) \ltimes \Sigma D \xrightarrow{J'_k} \Omega\Sigma X \ltimes (X^{\wedge k} \ltimes \Sigma D) \xrightarrow{1\ltimes\overline{\mathfrak{c}}'_k} \Omega\Sigma X \ltimes (\Omega\Sigma X \ltimes E)$$

is homotopic to $j_{k-1} \ltimes (j \circ \delta)$;

(d) *the composite*

$$J_{k-1}(X) \ltimes \Sigma D \xrightarrow{J'_k} \Omega\Sigma X \ltimes (X^{\wedge k} \ltimes \Sigma D) \xrightarrow{1\ltimes\overline{\mathfrak{c}}'_k} \Omega\Sigma X \ltimes (\Omega\Sigma X \ltimes E) \xrightarrow{\overline{a}\circ(1\ltimes\overline{a})} E$$

is homotopic to $\overline{a} \circ (1 \ltimes \overline{a}) \circ \overline{J}_k$.

PROOF. First, if $k = 1$ then it has already been observe that observe that $\overline{\mathfrak{c}}'_1 \circ i = \overline{\mathfrak{c}}_1$. For $k \geq 2$ observe that $\overline{\mathfrak{c}}'_k = (\overline{\mathfrak{c}}_k \perp j\circ\delta)\circ(q+\pi) = (\overline{\mathfrak{c}}_k \circ q) + (j \circ \delta \circ \pi)$. Observe also that $q \circ i$ is homotopic to the identity map on $X^{\wedge k} \wedge \Sigma D$ while $\pi \circ i$ is null homotopic. As Σi is a suspension it distributes on the right so we therefore obtain

$$\Sigma\overline{\mathfrak{c}}'_k \circ \Sigma i \simeq \Sigma\overline{\mathfrak{c}}_k \circ \Sigma q \circ \Sigma i + \Sigma j \circ \Sigma\delta \circ \Sigma\pi \circ \Sigma i \simeq\simeq \Sigma\overline{\mathfrak{c}}_k + * \simeq \Sigma\overline{\mathfrak{c}}_k,$$

proving part (a).

Second, if $k = 1$ the definition of $\overline{\mathfrak{c}}'_1$ as $E \ltimes \delta$ and the naturality of j implies that $\overline{\mathfrak{c}}'_1 \circ j = (E \ltimes \delta) \circ j \simeq j \circ \delta$. For $k \geq 2$ observe that $q \circ j$ is null homotopic while $\pi \circ j$ is homotopic to the identity map on ΣD. As $\Sigma D \xrightarrow{j} X^{\wedge k} \ltimes \Sigma D$ is a suspension it distributes on the right so we obtain

$$\overline{\mathfrak{c}}'_k \circ j = ((\overline{\mathfrak{c}}_k \circ q) + (j \circ \delta \circ \pi)) \circ j \simeq (\overline{\mathfrak{c}}_k \circ q \circ j) + (j \circ \delta \circ \pi \circ j) \simeq * + j \circ \delta,$$

proving part (b).

Third, by definition, $J'_k = j_{k-1} \ltimes j$ so by part (b) we have

$$(1 \ltimes \overline{\mathfrak{c}}'_k) \circ J'_k = (1 \ltimes \overline{\mathfrak{c}}'_k) \circ (j_{k-1} \ltimes j) = (j_{k-1} \ltimes (\overline{\mathfrak{c}}'_k \circ j)) = j_{k-1} \ltimes (j \circ \delta),$$

proving part (c).

Fourth, since $\overline{a}$ is the quotient of the homotopy action $\Omega\Sigma X \times E \xrightarrow{a} E$, which restricts to the identity map on E, the composite $E \xrightarrow{j} \Omega\Sigma X \ltimes E \xrightarrow{\overline{a}} E$ is homotopic to the identity map on E. Using this and part (c) we obtain

$$\overline{a} \circ (1 \ltimes \overline{a}) \circ (1 \ltimes \overline{\mathfrak{c}}'_k) \circ J'_k \simeq \overline{a} \circ (1 \ltimes \overline{a}) \circ (j_{k-1} \ltimes (j \circ \delta)) \simeq \overline{a} \circ (j_{k-1} \ltimes \delta).$$

On the other hand, by definition, $\overline{J}_k = j \circ (j_{k-1} \ltimes \delta)$, so using the naturality of j we have

$$\overline{a} \circ (1 \ltimes \overline{a}) \circ \overline{J}_k = \overline{a} \circ (1 \ltimes \overline{a}) \circ j \circ (j_{k-1} \ltimes \delta) \simeq \overline{a} \circ j \circ \overline{a} \circ (j_{k-1} \ltimes \delta) \simeq \overline{a} \circ (j_{k-1} \ltimes \delta).$$

Thus $\overline{a} \circ (1 \ltimes \overline{a}) \circ (1 \ltimes \overline{\mathfrak{c}}'_k) \circ J'_k \simeq \overline{a} \circ (1 \ltimes \overline{a}) \circ \overline{J}_k$, proving part (d). □

Recall from Chapter 3 that two maps f and g are congruent if $\Sigma f \simeq \Sigma g$.

LEMMA 5.6. *The composite*

$$(\Omega\Sigma X \ltimes (X^{\wedge k} \wedge \Sigma D)) \vee J_{k-1}(X) \ltimes \Sigma D \xrightarrow{e_1 \perp J'_k} \Omega\Sigma X \ltimes (X^{\wedge k} \ltimes \Sigma D) \xrightarrow{1 \ltimes \overline{\mathfrak{c}}'_k}$$
$$\Omega\Sigma X \ltimes (\Omega\Sigma X \ltimes E) \xrightarrow{\overline{a} \circ (1 \ltimes \overline{a})} E$$

is congruent to $\overline{a} \circ (1 \ltimes \overline{a}) \circ ((1 \ltimes \overline{\mathfrak{c}}_k) \perp \overline{J}_k)$ *(the map appearing in Proposition* 4.5*).*

PROOF. It is equivalent to prove the statement when restricted to each of the wedge summands $\Omega\Sigma X \ltimes (X^{\wedge k} \wedge \Sigma D)$ and $J_{k-1}(X) \ltimes \Sigma D$. By definition, e_1 is the map $\Omega\Sigma X \ltimes (X^{\wedge k} \wedge \Sigma D) \xrightarrow{1 \ltimes i} \Omega\Sigma X \ltimes (X^{\wedge k} \ltimes \Sigma D)$ and by Lemma 5.5 (a), $\overline{\mathfrak{c}}'_k \circ i$ is congruent to $\overline{\mathfrak{c}}_k$. Therefore $(1 \ltimes \overline{\mathfrak{c}}'_k) \circ e_1 = (1 \ltimes \overline{\mathfrak{c}}'_k)(1 \ltimes i)$ is congruent to $1 \ltimes \overline{\mathfrak{c}}_k$. Thus $\overline{a} \circ (1 \ltimes \overline{a}) \circ (1 \ltimes \overline{\mathfrak{c}}'_k) \circ e_1$ is congruent to $\overline{a} \circ (1 \ltimes \overline{a}) \circ (1 \ltimes \overline{\mathfrak{c}}_k)$.

On the other hand, by Lemma 5.5 (d), $\overline{a} \circ (1 \ltimes \overline{a}) \circ (1 \ltimes \overline{\mathfrak{c}}'_k) \circ J'_k \simeq \overline{a} \circ (1 \ltimes \overline{a}) \circ \overline{J}_k$. As homotopic maps are congruent, the lemma follows. □

Putting things together to this point gives the following.

PROPOSITION 5.7. *Suppose that there is a homotopy fibration sequence*

$$\Omega\Sigma X \xrightarrow{\partial} E \xrightarrow{p} Y \xrightarrow{h} \Sigma X$$

and a map $\Sigma A \xrightarrow{f} Y$ *that lifts through* p *to* $\Sigma A \xrightarrow{g} E$*, together satisfying the following properties:*

(a) *Ωh has a right homotopy inverse;*
(b) *there is a map $\delta\colon \Sigma D \longrightarrow E$ such that the composite*

$$\Omega\Sigma X \ltimes \Sigma D \xrightarrow{1 \ltimes \delta} \Omega\Sigma X \ltimes E \xrightarrow{\overline{a}} E$$

is a homotopy equivalence;

(c) *g factors as a composite*

$$g\colon \Sigma A \xrightarrow{\ell} X^{\wedge k} \ltimes \Sigma D \xrightarrow{\bar{\mathfrak{c}}'_k} \Omega\Sigma X \ltimes E \xrightarrow{\bar{a}} E$$

for some map ℓ;

(d) *the composite $\Sigma^2 A \xrightarrow{\Sigma\ell} \Sigma(X^{\wedge k} \ltimes \Sigma D) \xrightarrow{\Sigma\pi} \Sigma D$ is null homotopic.*

Then there is a homotopy commutative square

$$\begin{array}{ccc}
\Sigma(\Omega\Sigma X \ltimes \Sigma A) & \overline{\hspace{4em}} & \Sigma(\Omega\Sigma X \ltimes \Sigma A) \\
\Big\downarrow{\scriptstyle \Sigma\kappa_1 + *} & & \Big\downarrow{\scriptstyle \Sigma\theta} \\
\Sigma(\Omega\Sigma X \ltimes (X^{\wedge k} \wedge \Sigma D)) \vee \Sigma J_{k-1}(X) \ltimes \Sigma D & \xrightarrow{\Sigma\epsilon} & \Sigma E
\end{array}$$

where ϵ is a homotopy equivalence.

Proof. By hypothesis (c), g factors as $\Sigma A \xrightarrow{\ell} X^{\wedge k} \ltimes \Sigma D \xrightarrow{\bar{\mathfrak{c}}'_k} \Omega\Sigma X \ltimes E \xrightarrow{\bar{a}} E$. Taking the half-smash with the identity map on $\Omega\Sigma X$ then gives the commutativity of the left rectangle in the diagram

$$\begin{array}{ccccccc}
\Omega\Sigma X \ltimes \Sigma A & \overline{\hspace{4em}} & & \overline{\hspace{4em}} & \Omega\Sigma X \ltimes \Sigma A & & \\
\Big\downarrow{\scriptstyle (1\ltimes\ell)} & & & & \Big\downarrow{\scriptstyle 1\ltimes g} & \searrow{\scriptstyle \theta} & \\
\Omega\Sigma X \ltimes (X^{\wedge k} \ltimes \Sigma D)) & \xrightarrow{1\ltimes\bar{\mathfrak{c}}'_k} & \Omega\Sigma X \ltimes (\Omega\Sigma X \ltimes E) & \xrightarrow{1\ltimes\bar{a}} & \Omega\Sigma X \ltimes E & \xrightarrow{\bar{a}} & E.
\end{array}$$

The right triangle homotopy commutes by (2.6). Consider the diagram

$$\begin{array}{ccccc}
\Sigma(\Omega\Sigma X \ltimes \Sigma A) & \overline{\hspace{4em}} & \Sigma(\Omega\Sigma X \ltimes \Sigma A) & \overline{\hspace{2em}} & \Sigma(\Omega\Sigma X \ltimes \Sigma A) \\
\Big\downarrow{\scriptstyle \Sigma\kappa_1 + *} & & \Big\downarrow{\scriptstyle \Sigma(1\ltimes\ell)} & & \Big\downarrow{\scriptstyle \Sigma\theta} \\
\Sigma(\Omega\Sigma X \ltimes (X^{\wedge k} \wedge \Sigma D)) \vee \Sigma J_{k-1}(X) \ltimes \Sigma D & \xrightarrow{\Sigma e_1 \perp \Sigma J'_k} & \Sigma(\Omega\Sigma X \ltimes (X^{\wedge k} \ltimes \Sigma D)) & \xrightarrow{\Sigma\alpha} & \Sigma E
\end{array}$$

where $\alpha = \bar{a} \circ (1 \ltimes \bar{a}) \circ (1 \ltimes \bar{\mathfrak{c}}'_k)$. The left square homotopy commutes by Lemma 5.4 and the right square is the suspension of the previous diagram. Consider the composite along the bottom row and the string of identifications:

$$\begin{aligned}
\Sigma\alpha \circ (\Sigma e_1 \perp \Sigma J'_k) &= \Sigma\bar{a} \circ \Sigma(1 \ltimes \bar{a}) \circ \Sigma(1 \ltimes \bar{\mathfrak{c}}'_k) \circ (\Sigma e_1 \perp \Sigma J'_k) \\
&\simeq \Sigma\bar{a} \circ \Sigma(1 \ltimes \bar{a}) \circ (\Sigma(1 \ltimes \bar{\mathfrak{c}}_k) \perp \Sigma\overline{J}_k) \\
&= \Sigma\bar{a} \circ (1 \ltimes \bar{a}) \circ (1 \ltimes \bar{\mathfrak{c}}_k \perp \overline{J}_k).
\end{aligned}$$

The first equality is from the definition of α, the second is from Lemma 5.6 and the third is just pulling out a suspension coordinate. Proposition 4.5 shows that $\bar{a} \circ (1 \ltimes \bar{a}) \circ (1 \ltimes \bar{\mathfrak{c}}_k \perp \overline{J}_k)$ is a homotopy equivalence. Therefore $\Sigma\alpha \circ (\Sigma e_1 \perp \Sigma J'_k)$ is a homotopy equivalence. Taking $\epsilon = \alpha \circ (e_1 \perp J'_k)$ then gives the asserted homotopy commutative diagram and homotopy equivalence. □

The homotopy commutative diagram in Proposition 5.7 is the suspension of the diagram obtained in the proof of Theorem 4.6. Let $\overline{C}$ be the homotopy cofibre of the composite

$$\Sigma A \xrightarrow{\ell} X^{\wedge k} \ltimes \Sigma D \xrightarrow{q} X^{\wedge k} \wedge \Sigma D.$$

Then the homotopy cofibre of the map $\Sigma\kappa_1 + *$ in Proposition 5.7 is

$$\Sigma(\Omega\Sigma X \ltimes \overline{C}) \vee \Sigma J_{k-1}(X) \ltimes \Sigma D.$$

The homotopy cofibre of the map θ in Proposition 5.7 is E'. Therefore the homotopy commutativity of the diagram in the proposition implies that there is an induced map of cofibres

$$\psi\colon \Sigma(\Omega\Sigma X \ltimes \overline{C}) \vee \Sigma J_{k-1}(X) \ltimes \Sigma D \longrightarrow \Sigma E'$$

and the fact that $\Sigma\epsilon$ is a homotopy equivalence implies that ψ induces an isomorphism in homology by the five-lemma and so is a homotopy equivalence by Whitehead's Theorem. This gives a description of the homotopy type of $\Sigma E'$. However, we want to identify the homotopy type of E'. To do this an extra hypothesis is necessary.

THEOREM 5.8. *Suppose that there is a homotopy fibration sequence* $\Omega\Sigma X \xrightarrow{\partial} E \xrightarrow{p} Y \xrightarrow{h} \Sigma X$ *and a map* $\Sigma A \xrightarrow{f} Y$ *that lifts through* p *to* $\Sigma A \xrightarrow{g} E$*, together satisfying the following properties:*

(a) Ωh *has a right homotopy inverse;*

(b) *there is a map* $\delta\colon \Sigma D \longrightarrow E$ *such that the composite*

$$\Omega\Sigma X \ltimes \Sigma D \xrightarrow{1\ltimes\delta} \Omega\Sigma X \ltimes E \xrightarrow{\overline{a}} E$$

is a homotopy equivalence;

(c) g *factors as a composite*

$$g\colon \Sigma A \xrightarrow{\ell} X^{\wedge k} \ltimes \Sigma D \xrightarrow{\overline{\mathfrak{c}}'_k} \Omega\Sigma X \ltimes E \xrightarrow{\overline{a}} E$$

for some map ℓ*;*

(d) *the composite* $\Sigma^2 A \xrightarrow{\Sigma\ell} \Sigma(X^{\wedge k} \ltimes \Sigma D) \xrightarrow{\Sigma\pi} \Sigma D$ *is null homotopic;*

(e) *the composite* $\Sigma A \xrightarrow{\ell} X^{\wedge k} \ltimes \Sigma D \xrightarrow{q} X^{\wedge k} \wedge \Sigma D$ *has a left homotopy inverse.*

Then if $\overline{C}$ *is the homotopy cofibre of* $q \circ \ell$ *there is a homotopy equivalence*

$$E' \simeq (\Omega\Sigma X \ltimes \overline{C}) \vee (J_{k-1}(X) \ltimes \Sigma D).$$

PROOF. By Proposition 5.7, hypotheses (a) through (d) imply that there is a homotopy cofibration diagram

$$\begin{array}{ccc}
\Sigma(\Omega\Sigma X \ltimes \Sigma A) & = & \Sigma(\Omega\Sigma X \ltimes \Sigma A) \\
\downarrow{\scriptstyle \Sigma\kappa_1 + *} & & \downarrow{\scriptstyle \Sigma\theta} \\
\Sigma(\Omega\Sigma X \ltimes (X^{\wedge k} \wedge \Sigma D)) \vee (\Sigma J_{k-1}(X) \ltimes \Sigma D) & \xrightarrow{\Sigma\epsilon} & \Sigma E \\
\downarrow{\scriptstyle \Sigma\lambda\vee 1} & & \downarrow{\scriptstyle \Sigma\eta} \\
\Sigma(\Omega\Sigma X \ltimes \overline{C}) \vee (\Sigma J_{k-1}(X) \ltimes \Sigma D) & \xrightarrow{\psi} & \Sigma E'
\end{array} \tag{5.3}$$

where ϵ is a homotopy equivalence, λ is the map to the homotopy cofibre of κ_1, η is the map to the homotopy cofibre of θ, and ψ is an induced map of cofibres. As ϵ is a homotopy equivalence the five-lemma implies that ψ induces an isomorphism in homology and so is a homotopy equivalence by Whitehead's Theorem.

We wish to show that $\psi \simeq \Sigma\psi'$. Write $\psi = \psi_1 \perp \psi_2$ where ψ_1 and ψ_2 are the restrictions of ψ to $\Sigma(\Omega\Sigma X \ltimes \overline{C})$ and $\Sigma J_{k-1}(X) \ltimes \Sigma D$ respectively. Similarly write $\epsilon = \epsilon_1 \perp \epsilon_2$ where ϵ_1 and ϵ_2 are the restrictions of ϵ to $\Omega\Sigma X \ltimes (X^{\wedge k} \wedge \Sigma D)$ and $J_{k-1}(X) \ltimes \Sigma D$ respectively. Observe that the bottom square in (5.3) implies that

$\psi_2 = \Sigma\eta \circ \Sigma\epsilon_2$. In particular, ψ_2 is a suspension. Next, consider the homotopy cofibration $\Sigma A \xrightarrow{q\circ\ell} X^{\wedge k} \wedge \Sigma D \xrightarrow{\mu} \overline{C}$. By hypothesis (e), $q \circ \ell$ has a left homotopy inverse. As $X^{\wedge k} \wedge \Sigma D$ is a suspension this implies that the homotopy cofibration splits to give a homotopy equivalence

$$X^{\wedge k} \wedge \Sigma D \simeq \Sigma A \vee \overline{C}.$$

In particular, μ has a right homotopy inverse $\nu\colon \overline{C} \longrightarrow X^{\wedge k} \wedge \Sigma D$. Observe that by (5.2) and the definition of f_1 we have $\kappa_1 = f_1 \circ (1 \ltimes \ell) = (1 \ltimes q) \circ (1 \ltimes \ell)$. Therefore λ can be chosen to be $1 \ltimes \mu$ and so has $1 \ltimes \nu$ as a right homotopy inverse. Hence the bottom square in (5.3) implies that ψ_1 is homotopic to $\Sigma\eta \circ \Sigma\epsilon_1 \circ \Sigma\nu$. In particular, ψ_1 is a suspension. Hence ψ is a suspension, $\psi = \Sigma\psi'$. Since ψ is a homotopy equivalence, it induces an isomorphism in homology. Therefore so does ψ'. Since ΣD and $\overline{C}$ are simply-connected, so is $(\Omega\Sigma X \ltimes \overline{C}) \vee (J_{k-1}(X) \ltimes \Sigma D)$. Thus ψ' induces an isomorphism in homology between simply-connected spaces and so is a homotopy equivalence by Whitehead's Theorem. That is, there is a homotopy equivalence $E' \simeq (\Omega\Sigma X \ltimes \overline{C}) \vee (J_{k-1}(X) \ltimes \Sigma D)$. □

CHAPTER 6

Applying Theorem 5.8

In this section examples are given of Theorem 5.8 in action. This begins with a general example in Proposition 6.4 which will then lead to several more specific families of examples. We first need a general lemma.

For a space X, let

$$E\colon X \longrightarrow \Omega\Sigma X$$

be the suspension, which is adjoint to the identity map on ΣX. Given pointed, path-connected spaces $X_1, \ldots, X_m$, for $1 \leq s \leq m$ let

$$i_s\colon \Sigma X_s \longrightarrow \bigvee_{i=1}^{m} \Sigma X_i$$

be the inclusion of the sth-wedge summand. Let

$$I\colon \bigvee_{i=2}^{m} \Sigma X_i \longrightarrow \bigvee_{i=1}^{m} \Sigma X_i$$

be the inclusion, and note that $I = i_2 \perp \cdots \perp i_m$.

LEMMA 6.1. *Let $X_1, \ldots, X_m$ be pointed, path-connected spaces. Let*

$$q_1\colon \bigvee_{i=1}^{m} \Sigma X_i \longrightarrow \Sigma X_1$$

be the pinch map onto the first wedge summand and let E be its homotopy fibre. Then the following hold:

(a) *there is a map $g\colon \bigvee_{i=2}^{m} \Sigma X_i \longrightarrow E$ which lifts I through p;*
(b) *the composite*

$$\Omega\Sigma X_1 \ltimes (\bigvee_{i=2}^{m} \Sigma X_i) \xrightarrow{1\ltimes g} \Omega\Sigma X_1 \ltimes E \xrightarrow{\overline{a}} E$$

is a homotopy equivalence;
(c) *the composite*

$$X_1 \wedge (\bigvee_{i=2}^{m} \Sigma X_i) \xrightarrow{i} X_1 \ltimes (\bigvee_{i=2}^{m} \Sigma X_i) \xrightarrow{E\ltimes 1} \Omega\Sigma X_1 \ltimes (\bigvee_{i=2}^{m} \Sigma X_i) \xrightarrow{1\ltimes g} \Omega\Sigma X_1 \ltimes E \xrightarrow{\overline{a}} E$$

is a lift of the Whitehead product $[i_1, i_2] \perp \cdots \perp [i_1, i_m]$ through p.

PROOF. Let $X = X_1$ and $Y = \bigvee_{i=2}^{m} X_i$, so that $\bigvee_{i=1}^{m} \Sigma X_i = \Sigma X \vee \Sigma Y$. To avoid overlapping notation, let $i_L\colon \Sigma X \longrightarrow \Sigma X \vee \Sigma Y$ and $i_R\colon \Sigma Y \longrightarrow \Sigma X \vee \Sigma Y$ be the inclusions of the left and right wedge summands respectively. Since $q_1 \circ i_R$

is null homotopic, there is a map $g\colon \Sigma Y \longrightarrow E$ that lifts the inclusion through p. By Example 3.6 the composite

$$\Omega\Sigma X \ltimes \Sigma Y \xrightarrow{1\ltimes g} \Omega\Sigma X \ltimes E \xrightarrow{\overline{a}} E$$

is a homotopy equivalence. This proves parts (a) and (b). By Lemma 3.14 there is a homotopy commutative diagram

$$\begin{array}{ccc} X \wedge \Sigma Y & \xrightarrow{\eth_1} & E \\ & \searrow^{[i_L, i_R]} & \downarrow{\scriptstyle p} \\ & & \Sigma X \vee \Sigma Y. \end{array}$$

By Lemma 3.12, $\eth_1 = d_1$, where d_1 is the composite $X \wedge \Sigma Y \xrightarrow{c_1} \Omega\Sigma X \ltimes \Sigma Y \xrightarrow{1\ltimes g} \Omega\Sigma X \ltimes E \xrightarrow{\overline{a}} E$ and c_1 is the composite $X \wedge \Sigma Y \xrightarrow{i} X \ltimes \Sigma Y \xrightarrow{E\ltimes 1} \Omega\Sigma X \ltimes \Sigma Y$. Recalling that $X = X_1$ and $Y = \bigvee_{i=2}^m X_i$, we have $i_R = i_1$ and $i_L = i_2 \perp \cdots \perp i_m$. The linearity of the Whitehead product therefore implies that

$$[i_L, i_R] \simeq [i_1, i_2] \perp \cdots \perp [i_1, i_m].$$

Thus $\overline{a} \circ (1 \ltimes g) \circ (E \ltimes 1) \circ i$ is a lift of $[i_1, i_2] \perp \cdots \perp [i_1, i_m]$ through p, proving part (c). □

Parts (b) and (c) of Lemma 6.1 have the following corollaries.

COROLLARY 6.2. *Let $B \xrightarrow{\alpha} \bigvee_{i=2}^m \Sigma X_i$ be a map. A lift of the composite $B \xrightarrow{\alpha} \bigvee_{i=2}^m \Sigma X_i \xrightarrow{I} \bigvee_{i=1}^m \Sigma X_i$ through p is given by*

$$B \xrightarrow{\alpha} \bigvee_{i=2}^m \Sigma X_i \xrightarrow{j} X_1 \ltimes (\bigvee_{i=2}^m \Sigma X_i) \xrightarrow{E\ltimes 1} \Omega\Sigma X_1 \ltimes (\bigvee_{i=2}^m \Sigma X_i) \xrightarrow{1\ltimes g} \Omega\Sigma X \ltimes E \xrightarrow{\overline{a}} E \square$$

COROLLARY 6.3. *The restriction of the composite in Lemma 6.1 (c) to $X_1 \wedge \Sigma X_t$ for some $2 \leq t \leq m$ is a lift of the Whitehead product $[i_1, i_t]$ through p.* □

PROPOSITION 6.4. *Let $X_1, \ldots, X_m$ be pointed, path-connected spaces and suppose that there is a homotopy cofibration $\Sigma A \xrightarrow{f} \bigvee_{i=1}^m \Sigma X_i \longrightarrow M$. Suppose that $f = f_1 + f_2$ where:*

- *$f_1 = \sum_{j=2}^m [i_1, i_j] \circ h_{1,j}$ for some maps $h_{1,j}\colon \Sigma A \longrightarrow \Sigma X_1 \wedge X_j$;*
- *there is at least one $t \in \{2, \ldots, m\}$ such that $\Sigma A \xrightarrow{h_{1,t}} \Sigma X_1 \wedge X_t$ has a left homotopy inverse;*
- *f_2 factors as $\Sigma A \xrightarrow{\gamma} \bigvee_{i=2}^m \Sigma X_i \xrightarrow{I} \bigvee_{i=1}^m \Sigma X_i$ for some map γ;*
- *$\Sigma\gamma$ is null homotopic.*

Let $h = \sum_{j=2}^m h_{1,j}$ and let $\overline{C}$ be the homotopy cofibre of $\Sigma A \xrightarrow{h} X_1 \wedge (\bigvee_{i=2}^m \Sigma X_i)$. Then the following hold:

(a) *there is a map $q'\colon M \longrightarrow \Sigma X_1$ extending q_1;*
(b) *there is a homotopy fibration*

$$(\Omega\Sigma X_1 \ltimes \overline{C}) \vee (\bigvee_{i=2}^m \Sigma X_i) \longrightarrow M \xrightarrow{q'} \Sigma X_1;$$

(c) *the homotopy fibration in part (b) splits after looping to give a homotopy equivalence*

$$\Omega M \simeq \Omega\Sigma X_1 \times \Omega\Big((\Omega\Sigma X_1 \ltimes \overline{C}) \vee (\bigvee_{i=2}^{m} \Sigma X_i)\Big).$$

PROOF. First observe that as f_1 factors through the Whitehead products $[i_1, i_j]$ and f_2 factors through I, the composite $\Sigma A \xrightarrow{f} \bigvee_{i=1}^{m} \Sigma X_i \xrightarrow{q_1} \Sigma X_1$ is null homotopic, so q_1 extends to a map $q' \colon M \longrightarrow \Sigma X_1$. This proves part (a).

To prove parts (b) and (c), Theorem 5.8 will be applied to the homotopy fibration $E \xrightarrow{p} \bigvee_{i=1}^{m} \Sigma X_i \xrightarrow{q_1} \Sigma X_1$ and the attaching map f for M. The hypotheses for that theorem need to be checked. In the notation of Proposition 5.8, let $D = \bigvee_{i=2}^{m} X_i$ and let $\Sigma D \xrightarrow{\delta} E$ be a lift of I through p.

STEP 1. The map i_1 is a right homotopy inverse for q_1, so hypothesis (a) in Theorem 5.8 is satisfied. With D and δ as above, the homotopy equivalence in Lemma 6.1 (b) implies that hypothesis (b) of Theorem 5.8 is satisfied.

STEP 2. For hypothesis (c) of Theorem 5.8 we need to choose a lift g of f through p. Let $\ell_{1,j}$ be the composite

$$\ell_{1,j} \colon \Sigma A \xrightarrow{h_{1,j}} \Sigma X_1 \wedge X_j \hookrightarrow X_1 \wedge (\bigvee_{i=2}^{m} \Sigma X_i) \xrightarrow{i} X_1 \ltimes (\bigvee_{i=2}^{m} \Sigma X_i).$$

Then by Corollary 6.3 the composite

$$\Sigma A \xrightarrow{\ell_{1,j}} X_1 \ltimes (\bigvee_{i=2}^{m} \Sigma X_i) \xrightarrow{E\ltimes\delta} \Omega\Sigma X_1 \ltimes E \xrightarrow{\overline{a}} E$$

is a lift of $[i_1, i_j] \circ h_{1,j}$ through p. Let $\ell_1 = \sum_{j=2}^{m} \ell_{1,j}$. Then the composite

$$g_1 \colon \Sigma A \xrightarrow{\ell_1} X_1 \ltimes (\bigvee_{i=2}^{m} \Sigma X_i) \xrightarrow{E\ltimes\delta} \Omega\Sigma X_1 \ltimes E \xrightarrow{\overline{a}} E$$

is a lift of f_1 through p. Let ℓ_2 be the composite

$$\ell_2 \colon \Sigma A \xrightarrow{\gamma} \bigvee_{i=2}^{m} \Sigma X_i \xrightarrow{j} X_1 \ltimes (\bigvee_{i=2}^{m} \Sigma X_i).$$

Then by Corollary 6.2 the composite

$$g_2 \colon \Sigma A \xrightarrow{\ell_2} X_1 \ltimes (\bigvee_{i=2}^{m} \Sigma X_i) \xrightarrow{E\ltimes\delta} \Omega\Sigma X_1 \ltimes E \xrightarrow{\overline{a}} E$$

is a lift of f_2 through p. Thus if

$$\ell \colon \Sigma A \longrightarrow X_1 \ltimes (\bigvee_{i=2}^{m} \Sigma X_i)$$

is $\ell_1 + \ell_2$ and

$$g \colon \Sigma A \xrightarrow{\ell} X_1 \ltimes (\bigvee_{i=2}^{m} \Sigma X_i) \xrightarrow{E\ltimes\delta} \Omega\Sigma X_1 \ltimes E \xrightarrow{\overline{a}} E$$

is $g_1 + g_2$ then g is a lift of f through p. By definition, the map $\mathfrak{c}_1'$ in Chapter 5 equals $E \ltimes 1$, so the map g satisfies hypothesis (c) of Theorem 5.8.

STEP 3. Consider the composite

$$\Sigma A \xrightarrow{\ell} X_1 \ltimes (\bigvee_{i=2}^{m} \Sigma X_i) \xrightarrow{\pi} \bigvee_{i=2}^{m} \Sigma X_i.$$

By definition, $\ell_{1,j}$ factors through i and $\pi \circ i$ is null homotopic. Therefore $\pi \circ \ell_{1,j}$ is null homotopic for each $2 \leq j \leq m$. As $\ell_1 = \sum_{j=2}^{m} \ell_{1,j}$, we obtain a null homotopy for $\pi \circ \ell_1$. By definition, $\ell_2 = j \circ \gamma$ and $\pi \circ j$ is the identity map, so $\pi \circ \ell_2 = \gamma$. By hypothesis, $\Sigma\gamma$ is null homotopic, and therefore so is $\Sigma(\pi \circ \ell_2)$. As $\ell = \ell_1 + \ell_2$, we obtain a null homotopy for $\Sigma(\pi \circ \ell)$. This fulfils hypothesis (d) of Theorem 5.8.

STEP 4. Consider the composition

$$\Sigma A \xrightarrow{\ell} X_1 \ltimes (\bigvee_{i=2}^{m} \Sigma X_i) \xrightarrow{q} X_1 \wedge (\bigvee_{i=2}^{m} \Sigma X_i).$$

By definition, $\ell_2 = j \circ \gamma$ and $q \circ j$ is null homotopic, so $q \circ \ell_2$ is null homotopic. Therefore, as $\ell = \ell_1 + \ell_2$, we have $q \circ \ell \simeq q \circ \ell_1$. On the other hand, by definition, $\ell_{1,t}$ factors through i and $q \circ i$ is homotopic to the identity map. Therefore $q \circ \ell_{1,j}$ is homotopic to the composite

$$\Sigma A \xrightarrow{h_{1,j}} \Sigma X_1 \wedge X_j \hookrightarrow X_1 \wedge (\bigvee_{i=2}^{m} \Sigma X_i). \tag{6.1}$$

The sum of the inclusions $\Sigma X_1 \wedge X_j \longrightarrow X_1 \wedge (\bigvee_{i=2}^{m} \Sigma X_i)$ for $2 \leq j \leq m$ is homotopic to the identity map, so as $h = \sum_{j=2}^{m} h_{1,j}$ and $\ell = \sum_{j=2}^{m} \ell_{1,j}$ we have $q \circ \ell_1$ homotopic to h. Hence $q \circ \ell \simeq h$.

STEP 5. By hypothesis, there is a $t \in \{2, \ldots, m\}$ such that $\Sigma A \xrightarrow{h_{1,t}} \Sigma X_1 \wedge X_t$ has a left homotopy inverse $r\colon \Sigma X_1 \wedge X_t \longrightarrow \Sigma A$. Consider the composite

$$\Sigma A \xrightarrow{h} X_1 \wedge (\bigvee_{i=2}^{m} \Sigma X_i) \xrightarrow{1 \wedge q_t} X_1 \wedge \Sigma X_t \xrightarrow{r} \Sigma A \tag{6.2}$$

where q_t is the pinch map to the tth-wedge summand. Observe that (6.1) composed with $1 \wedge q_t$ is null homotopic if $j \neq t$ and is homotopic to $h_{1,t}$ if $j = t$. Thus in (6.2) the composite $(1 \wedge q_t) \circ h$ is homotopic to $h_{1,t}$. Hence as r is a left homotopy inverse for $h_{1,t}$, the composite (6.2) is homotopic to the identity map. In particular, h has a left homotopy inverse. That is, by Step 4, $q \circ \ell$ has a left homotopy inverse. This fulfils hypothesis (e) of Theorem 5.8.

STEP 6. As hypotheses (a) to (e) of Theorem 5.8 hold, applying the proposition immediately implies assertions (b) and (c), noting that by Step 3, $h = q \circ \ell$.

□

The homotopy decomposition of ΩM in Proposition 6.4 can be made more precise by identifying the homotopy type of $\overline{C}$. One hypothesis is that for some $t \in \{2, \ldots, m\}$ the map $\Sigma A \xrightarrow{h_{1,t}} \Sigma X_1 \wedge X_t$ has a left homotopy inverse. Let B be the homotopy cofibre of $h_{1,t}$. The left homotopy inverse for $h_{1,t}$ and the fact that $\Sigma X_1 \wedge X_t$ is a suspension implies that there is a homotopy equivalence

$$\Sigma X_1 \wedge X_t \simeq \Sigma A \vee B.$$

LEMMA 6.5. *In Proposition 6.4, there is a homotopy equivalence*

$$\overline{C} \simeq \Big(X_1 \wedge (\bigvee_{\substack{i=2 \\ i\neq t}} \Sigma X_i)\Big) \vee B.$$

PROOF. Let $q_t\colon \bigvee_{i=1}^m \Sigma X_i \longrightarrow X_t$ be the pinch map to the tth-wedge summand. Then $q_t \circ h = h_{1,t}$, so there is a homotopy cofibration diagram

$$\tag{6.3} \begin{array}{ccccc} & & X_1 \wedge (\bigvee_{\substack{i=2 \\ i\neq t}} \Sigma X_i) & = & X_1 \wedge (\bigvee_{\substack{i=2 \\ i\neq t}} \Sigma X_i) \\ & & \downarrow & & \downarrow \\ \Sigma A & \xrightarrow{h} & X_1 \wedge (\bigvee_{i=1}^m \Sigma X_i) & \longrightarrow & \overline{C} \\ \| & & \downarrow{\scriptstyle 1\wedge q_t} & & \downarrow \\ \Sigma A & \xrightarrow{h_{1,t}} & X_1 \wedge \Sigma X_t & \longrightarrow & B. \end{array}$$

The homotopy equivalence $\Sigma X_1 \wedge X_t \simeq \Sigma A \vee B$ splitting the homotopy cofibration along the bottom row implies that the map $\Sigma X_1 \wedge X_t \longrightarrow B$ has a right homotopy inverse $b\colon B \longrightarrow X_1 \wedge \Sigma X_t$. As i_t is a right homotopy inverse for q_t, we obtain a composite

$$B \xrightarrow{b} X_1 \wedge \Sigma X_t \xrightarrow{1\wedge i_t} X_1 \wedge (\bigvee_{i=1}^m \Sigma X_i) \longrightarrow \overline{C}$$

which, by the homotopy commutativity of the lower right square in (6.3), is a right homotopy inverse for the map $\overline{C} \longrightarrow B$. Thus the right column in (6.3) splits to give the asserted homotopy equivalence. □

A family of examples satisfying Proposition 6.4 is the following. In words it says that if there is a homotopy cofibration $S^{2n-1} \xrightarrow{f} \bigvee_{i=1}^m S^n \longrightarrow M$ where the attaching map f is: (i) a sum of Whitehead products, at least one of which is $\pm[i_1, i_t]$ for some $t \in \{2, \ldots, m\}$, and (ii) a map factoring through $\bigvee_{i=2}^m S^n$ that suspends trivially, then the homotopy type of ΩM can be precisely determined.

PROPOSITION 6.6. *Suppose that there is a homotopy cofibration*

$$S^{2n-1} \xrightarrow{f} \bigvee_{i=1}^m S^n \longrightarrow M.$$

Suppose that $f = f_1 + f_2$ where:

- *$f_1 = \sum_{j=2}^m d_j \cdot [i_1, i_j]$ for $d_j \in \mathbb{Z}$;*
- *there is at least one $t \in \{2, \ldots, m\}$ such that $d_t = \pm 1$;*
- *f_2 factors as $S^{2n-1} \xrightarrow{\gamma} \bigvee_{i=2}^m S^n \xrightarrow{I} \bigvee_{i=1}^m S^n$ for some map γ;*
- *$\Sigma\gamma$ is null homotopic.*

Then there is a homotopy fibration

$$(\Omega S^n \ltimes \overline{C}) \vee (\bigvee_{i=2}^m S^n) \longrightarrow M \xrightarrow{q'} S^n$$

where $\overline{C} \simeq S^{n-1} \wedge (\bigvee_{\substack{i=2 \\ i\neq t}}^{m} S^n)$*, and this homotopy fibration splits after looping to give a homotopy equivalence*

$$\Omega M \simeq \Omega S^n \times \Omega\Big((\Omega S^n \ltimes \overline{C}) \vee (\bigvee_{i=2}^{m} S^n)\Big).$$

PROOF. The existence of the homotopy fibration and the decomposition for ΩM will follow from Proposition 6.4 once the hypotheses on the attaching map f are shown to imply the hypotheses in the proposition. Observe that the map $\Sigma A \xrightarrow{h_{1,j}} \Sigma X_1 \wedge X_j$ in Proposition 6.4 in our case is of the form $S^{2n-1} \longrightarrow S^{2n-1}$ and so is a degree map, which has been labelled d_j. The condition that $d_t = \pm 1$ for some $t \in \{2, \ldots, m\}$ implies that the map $1 \wedge h_{1,t} \simeq 1 \wedge d_t$ is a homotopy equivalence, and so has a right homotopy inverse. The conditions on f_2 and γ are the same as in Proposition 6.4. The homotopy type of $\overline{C}$ follows from Lemma 6.5, noting that as $h_{1,t}$ is a homotopy equivalence its homotopy cofibre B is contractible. □

REMARK 6.7. Observe that the homotopy type of ΩM in Proposition 6.6 depends only on m and n. In particular, the map γ has no influence on the homotopy type.

COROLLARY 6.8. *In Proposition* 6.6 *the space* $(\Omega S^n \ltimes \overline{C}) \vee (\bigvee_{i=2}^{m} S^{n_i+1}))$ *is homotopy equivalent to a wedge W of spheres. In particular,*

$$\Omega M \simeq \Omega S^{n_1+1} \times \Omega W.$$

PROOF. It suffices to show that $\Omega S^{n_1+1} \ltimes \overline{C}$ is homotopy equivalent to a wedge of spheres. By Proposition 6.6, $\overline{C}$ is homotopy equivalent to a wedge of simply-connected spheres. In particular, $\overline{C} \simeq \Sigma C'$ where C' is a wedge of connected spheres. Therefore

$$\Omega S^n \ltimes C \simeq \Omega S^n \ltimes \Sigma C' \simeq (\Sigma\Omega S^n \wedge C') \vee \Sigma C'.$$

The James construction implies that $\Sigma\Omega S^n$ is homotopy equivalent to a wedge of spheres, and therefore so is $(\Sigma\Omega S^n) \wedge C'$. Hence $\Omega S^n \ltimes \overline{C}$ is homotopy equivalent to a wedge of spheres. □

We give two examples of Proposition 6.6. The first is not new, as it can be derived from the results in any one of [**BT1, BT2, BB**]. The second is new in general.

EXAMPLE 6.9. In Proposition 6.6, if the cofibration takes the form

$$S^{2n-1} \xrightarrow{f} \bigvee_{i=1}^{2m} S^n \longrightarrow M$$

where $f = [i_1, i_2] + [i_3, i_4] + \cdots + [i_{2m-1}, i_{2m}]$ then M is an $(n-1)$-connected $2n$-dimensional Poincaré Duality complex. In fact, it is the m-fold connected sum $(S^n \times S^n)^{\#m}$. Proposition 6.6 then gives a homotopy decomposition of ΩM.

EXAMPLE 6.10. Modifying the previous example, consider a homotopy cofibration

$$S^{2n-1} \xrightarrow{f'} \bigvee_{i=1}^{2m} S^n \longrightarrow M'$$

where $f' = [i_1, i_2] + [i_3, i_4] + \cdots + [i_{2m-1}, i_{2m}] + f''$. Here, f'' is a composite $f''\colon S^{2n-1} \xrightarrow{\gamma} \bigvee_{i=2}^{2m} S^n \xrightarrow{I} \bigvee_{i=1}^{2m} S^n$ with the property that $\Sigma\gamma$ is null homotopic. Possibly γ is a sum of more Whitehead products, possibly it is a class of finite order, or some combination of the two. Then M' may or may not be a Poincaré Duality complex but Proposition 6.6 still applies, giving a homotopy decomposition of $\Omega M'$. Note that the decompositions for $\Omega M'$ and the space ΩM in the previous example are identical. That is, while f'' may mean $M \not\simeq M'$, after looping we nevertheless have $\Omega M \simeq \Omega M'$.

Next, we consider the case when the spaces X_i in Proposition 6.4 are mod-p^r Moore spaces. The analogue of Proposition 6.6 involves mod-p^r Whitehead products rather than ordinary Whitehead products. Let $a\colon P^{m+1}(p^r) \longrightarrow Z$ and $b\colon P^{n+1}(p^r) \longrightarrow Z$ be maps. If $p^r \neq 2$, by Lemma 4.20 there is a map

$$u\colon P^{m+n+1}(p^r) \longrightarrow \Sigma P^m(p^r) \wedge P^n(p^r)$$

which has a left homotopy inverse. The *mod-p^r Whitehead product* is the composite

$$[a, b]_r\colon P^{m+n+1}(p^r) \xrightarrow{u} \Sigma P^m(p^r) \wedge P^n(p^r) \xrightarrow{[a,b]} Z$$

where $[a, b]$ is the usual Whitehead product.

LEMMA 6.11. *Suppose that for $n \geq 2$ there is a homotopy cofibration*

$$P^{2n+1}(p^r) \xrightarrow{f} \bigvee_{i=1}^{m} P^{n+1}(p^r) \longrightarrow M.$$

Suppose that $f = f_1 + f_2$ where:

- *$f_1 = \sum_{j=2}^{m} d_j \cdot [i_1, i_j]_r$ for $d_j \in \mathbb{Z}/p^r\mathbb{Z}$;*
- *there is at least one $t \in \{2, \ldots, m\}$ such that d_t is a unit in $\mathbb{Z}/p^r\mathbb{Z}$;*
- *f_2 factors as $P^{2n+1}(p^r) \xrightarrow{\gamma} \bigvee_{i=2}^{m} P^{n+1}(p^r) \xrightarrow{I} \bigvee_{i=1}^{m} P^{n+1}(p^r)$ for some map γ;*
- *$\Sigma\gamma$ is null homotopic.*

Then there is a homotopy fibration

$$(\Omega P^{n+1}(p^r) \ltimes \overline{C}) \vee (\bigvee_{i=2}^{m} P^{n+1}(p^r)) \longrightarrow M \xrightarrow{q'} P^{n+1}(p^r)$$

where $\overline{C} \simeq \left(P^n(p^r) \wedge (\bigvee_{\substack{i=2\\ i\neq t}}^{m} P^{n+1}(p^r))\right) \vee P^{2n}(p^r)$, *and this homotopy fibration splits after looping to give a homotopy equivalence*

$$\Omega M \simeq \Omega P^{n+1}(p^r) \times \Omega\Big((\Omega P^{n+1}(p^r) \ltimes \overline{C}) \vee (\bigvee_{i=2}^{m} P^{n+1}(p^r))\Big).$$

PROOF. The argument is just as for Proposition 6.6, but with the map $\Sigma A \xrightarrow{h_{1,j}} \Sigma X_1 \wedge X_j$ in Proposition 6.4 in this case being $P^{2n+1}(p^r) \xrightarrow{d_j \cdot u} \Sigma P^n(p^r) \wedge P^n(p^r)$. $\square$

REMARK 6.12. As for Proposition 6.6, the homotopy type of ΩM in Lemma 6.11 depends only on m and n, with the map γ having no influence on the homotopy type.

COROLLARY 6.13. *In Lemma* 6.11 *the space* $(\Omega P^{n+1}(p^r)\ltimes\overline{C})\vee(\bigvee_{i=2}^{m}P^{n+1}(p^r))$ *is homotopy equivalent to a wedge* W' *of mod-*p^r *Moore spaces. In particular,*

$$\Omega M\simeq\Omega P^{n+1}(p^r)\times\Omega W'.$$

PROOF. The argument is just as for Corollary 6.8 with appeals to Lemma 4.20 in order to decompose iterated smash products of mod-p^r Moore spaces into wedges of mod-p^r Moore spaces. □

CHAPTER 7

An application to Poincaré Duality spaces

In Lemma 6.11 a mod-p^r Moore space was attached to a wedge of mod-p^r Moore spaces. We want to next consider attaching a sphere to a wedge of mod-p^r Moore spaces. That is, we consider a homotopy cofibration of the form

$$S^{2n} \xrightarrow{f} \bigvee_{i=1}^{m} P^{n+1}(p^r) \longrightarrow M$$

and will assume throughout the section that p is odd and $n \geq 2$. Such cofibrations are particularly interesting because for certain maps f the space M is an $(n-1)$-connected $(2n+1)$-dimensional Poincaré Duality complex that is rationally equivalent to S^{2n+1}. A highlight of this section is the proof of Theorem 1.3.

The distinction between attaching a Moore space and attaching a sphere is large in the sense that we can no longer appeal to Proposition 6.4 or even to Theorem 5.8. Instead, we have to go back to the inner workings of the proof of Theorem 5.8 and make a modification that is specific to this case. This requires some initial lemmas.

LEMMA 7.1. *Suppose that there is a map $A \xrightarrow{v} X^{\wedge k} \wedge D$ such that $X \wedge \Sigma A \xrightarrow{1\wedge\Sigma v} X \wedge (X^{\wedge k} \wedge \Sigma D)$ has a left homotopy inverse. Then the map $\Omega\Sigma X \wedge \Sigma A \xrightarrow{1\wedge\Sigma v} \Omega\Sigma X \wedge (X^{\wedge k} \wedge \Sigma D)$ has a left homotopy inverse.*

PROOF. It will be convenient to write the identity map on a space Z as 1_Z. Let $u\colon X \wedge (X^{\wedge k} \wedge \Sigma D) \longrightarrow X \wedge \Sigma A$ be a left homotopy inverse for the map $1_X \wedge \Sigma v$. Then for each $t \geq 1$ the composite

$$X^{\wedge t} \wedge \Sigma A \xrightarrow{1_{X^{\wedge t}}\wedge\Sigma v} X^{\wedge t} \wedge (X^{\wedge k} \wedge \Sigma D) \xrightarrow{1_{X^{\wedge t}}\wedge u} X^{\wedge t-1} \wedge \Sigma A$$

is homotopic to the identity map. Consider the diagram

$$\begin{array}{ccccc} \Omega\Sigma X \wedge \Sigma A & \xrightarrow{1_{\Omega\Sigma X}\wedge\Sigma v} & \Omega\Sigma X \wedge \Sigma D & & \\ \downarrow\simeq & & \downarrow\simeq & & \\ \bigvee_{t=1}^{\infty} X^{\wedge t} \wedge \Sigma A & \xrightarrow{\bigvee_{t=1}^{\infty} 1_{X^{\wedge t}}\wedge\Sigma v} & \bigvee_{t=1}^{\infty} X^{\wedge t} \wedge (X^{\wedge k} \wedge \Sigma D) & \xrightarrow{\bigvee_{t=1}^{\infty} 1_{X^{\wedge t}}\wedge u} & \bigvee_{t=1}^{\infty} X^{\wedge t} \wedge \Sigma A. \end{array}$$

The right square homotopy commutes by the naturality of the James splitting of $\Sigma\Omega\Sigma X$, where we have used the fact that Σv is a suspension to rewrite $1_{\Omega\Sigma X} \wedge \Sigma v$ as $1_{\Sigma\Omega\Sigma X} \wedge v$. Observe that the bottom row is homotopic to the identity map since $(1_{X^{\wedge t}} \wedge u) \circ (1_{X^{\wedge t}} \wedge \Sigma v)$ is homotopic to the identity map for each $t \geq 1$. Therefore the homotopy commutativity of the diagram implies that $1_{\Omega\Sigma X} \wedge \Sigma v$ has a left homotopy inverse. □

EXAMPLE 7.2. The relevance of Lemma 7.1 is as follows. Let v be the composite

$$v\colon S^{2n-1} \longrightarrow P^{2n}(p^r) \xrightarrow{u} P^n(p^r) \wedge P^n(p^r)$$

where the left map is the inclusion of the bottom cell and u is the inclusion of the top dimensional Moore space in the homotopy decomposition of $P^n(p^r) \wedge P^n(p^r)$ in Lemma 4.20. In particular, Σv does not have a left homotopy inverse. However, Lemma 4.20 implies that

$$P^n(p^r) \wedge S^{2n} \xrightarrow{1\wedge\Sigma v} P^n(p^r) \wedge (\Sigma P^n(p^r) \wedge P^n(p^r))$$

does have a left homotopy inverse. Lemma 7.1 then implies that

$$\Omega P^{n+1}(p^r) \wedge S^{2n} \xrightarrow{1\wedge\Sigma v} \Omega P^{n+1}(p^r) \wedge (\Sigma P^n(p^r) \wedge P^n(p^r))$$

has a left homotopy inverse.

Next, we consider what will be the analogue of the map ℓ in Theorem 5.8.

LEMMA 7.3. *Suppose that there is a map $S^{2n} \xrightarrow{\ell} P^n(p^r) \ltimes (\bigvee_{i=2}^m P^{n+1}(p^r))$ which induces an inclusion in mod-p homology. If p is odd then the order of ℓ is p^r and ℓ factors as a composite*

$$S^{2n} \longrightarrow P^{2n+1}(p^r) \xrightarrow{\ell'} P^n(p^r) \ltimes (\bigvee_{i=2}^m P^{n+1}(p^r))$$

for some map ℓ'.

PROOF. By Lemmas 2.1 and 4.20 there are homotopy equivalences

$$\begin{aligned} P^n(p^r) \ltimes (\bigvee_{i=2}^m P^{n+1}(p^r)) &\simeq \Big(\bigvee_{i=2}^m P^n(p^r) \wedge P^{n+1}(p^r)\Big) \vee (\bigvee_{i=2}^m P^{n+1}(p^r)) \\ &\simeq \Big(\bigvee_{i=2}^m P^{2n+1}(p^r) \vee P^{2n}(p^r)\Big) \vee (\bigvee_{i=2}^m P^{n+1}(p^r)). \end{aligned}$$

Since ℓ induces an inclusion in mod-p homology, it must map into at least one of the $P^{2n+1}(p^r)$ wedge summands as the inclusion of the bottom cell (up to multiplication by a unit in $\mathbb{Z}/p^r\mathbb{Z}$). Therefore the order of ℓ is at least p^r. On the other hand, by [**CMN**], if p is odd then $\pi_k(P^s(p^r))$ is annihilated by p^r for any $k \leq 2s$. The Hilton-Milnor Theorem implies that any wedge $\bigvee_{j=1}^t P^{s_j}(p^r)$ with $n+1 \leq s_j \leq 2n+1$ for all $1 \leq j \leq t$ has the property that $\pi_k(\bigvee_{j=1}^t P^{s_j}(p^r))$ is annihilated by p^r for all $k \leq 2n+2$. Thus, in our case, $\pi_{2n}(P^n(p^r) \ltimes (\bigvee_{i=2}^m P^{n+1}(p^r)))$ is annihilated by p^r, so the order of ℓ is at most p^r. Hence the order of ℓ is exactly p^r. Consequently, ℓ extends to $P^{2n+1}(p^r) \xrightarrow{\ell'} P^n(p^r) \ltimes (\bigvee_{i=2}^m P^{n+1}(p^r))$ for some map ℓ'. □

Define the spaces $\widetilde{C}$ and $\overline{C}$ by the homotopy cofibration diagram

$$\begin{array}{ccccc} & & \bigvee_{i=2}^m P^{n+1}(p^r) & = & \bigvee_{i=2}^m P^{n+1}(p^r) \\ & & \downarrow & & \downarrow \\ S^{2n} & \xrightarrow{\ell} & P^n(p^r) \ltimes (\bigvee_{i=2}^m P^{n+1}(p^r)) & \longrightarrow & \widetilde{C} \\ \| & & \downarrow{\scriptstyle q} & & \downarrow \\ S^{2n} & \xrightarrow{q\circ\ell} & P^n(p^r) \wedge (\bigvee_{i=2}^m P^{n+1}(p^r)) & \longrightarrow & \overline{C}. \end{array} \tag{7.1}$$

Since ℓ induces an inclusion in mod-p homology so does $q \circ \ell$. In particular, there is a $t \in \{2, \ldots, m\}$ such that the composite

$$S^{2n} \xrightarrow{q \circ \ell} P^n(p^r) \wedge (\bigvee_{i=2}^{m} P^{n+1}(p^r)) \xrightarrow{1 \wedge q_t} P^n(p^r) \wedge P^{n+1}(p^r)$$

induces an injection in mod-p homology.

LEMMA 7.4. *The composite*

$$P^{2n+1}(p^r) \xrightarrow{\ell'} P^n(p^r) \ltimes (\bigvee_{i=2}^{m} P^{n+1}(p^r)) \xrightarrow{q} P^n(p^r) \wedge (\bigvee_{i=2}^{m} P^{n+1}(p^r)) \xrightarrow{1 \wedge q_t} P^n(p^r) \wedge P^{n+1}(p^r)$$

has a left homotopy inverse.

PROOF. The restriction of $(1 \wedge q_t) \circ q \circ \ell'$ to S^{2n} is $(1 \wedge q_t) \circ q \circ \ell$, which is an injection in mod-p homology. The action of the Bockstein then implies that $(1 \wedge q_t) \circ q \circ \ell'$ induces an injection in mod-p homology. By Lemma 4.20, $P^n(p^r) \wedge P^{n+1}(p^r) \simeq P^{2n+1}(p^r) \vee P^{2n}(p^r)$, so composing $(1 \wedge q_t) \circ q \circ \ell'$ with the pinch map to $P^{2n+1}(p^r)$ gives a self-map of $P^{2n+1}(p^r)$ which induces an isomorphism in mod-p homology, and hence in integral homology, and so is a homotopy equivalence by Whitehead's Theorem. □

LEMMA 7.5. *Suppose that there is a map $S^{2n} \xrightarrow{\ell} P^n(p^r) \ltimes (\bigvee_{i=2}^{m} P^{n+1}(p^r))$ which induces an inclusion in mod-p homology. If p is odd then the following hold:*

(a) *the homotopy cofibration $\bigvee_{i=2}^{m} P^{n+1}(p^r) \longrightarrow \widetilde{C} \longrightarrow \overline{C}$ in (7.1) splits to give a homotopy equivalence*

$$\widetilde{C} \simeq (\bigvee_{i=2}^{m} P^{n+1}(p^r)) \vee \overline{C};$$

(b) *there is a homotopy equivalence*

$$\Big(P^n(p^r) \wedge (\bigvee_{\substack{i=2 \\ i \neq t}}^{m} P^{n+1}(p^r))\Big) \vee \Big(S^{2n+1} \vee P^{2n}(p^r)\Big) \xrightarrow{\simeq} \overline{C}$$

where the map $S^{2n+1} \longrightarrow \overline{C}$ factors through the map $\widetilde{C} \longrightarrow \overline{C}$.

PROOF. The hypotheses imply that Lemma 7.3 holds. The factorization of ℓ through ℓ' in Lemma 7.3 implies that there is a homotopy cofibration diagram

$$\begin{array}{ccccc} S^{2n} & \longrightarrow & P^{2n+1}(p^r) & \longrightarrow & S^{2n+1} \\ \| & & \downarrow{\scriptstyle \ell'} & & \downarrow \\ S^{2n} & \xrightarrow{\ell} & P^n(p^r) \ltimes (\bigvee_{i=2}^{m} P^{n+1}(p^r)) & \longrightarrow & \widetilde{C} \\ & & \downarrow{\scriptstyle a} & & \downarrow{\scriptstyle b} \\ & & G & = & G \end{array} \tag{7.2}$$

which defines the space G and the maps a and b. By Lemma 7.4, ℓ' has a left homotopy inverse. Therefore as $P^n(p^r) \ltimes (\bigvee_{i=2}^{m} P^{n+1}(p^r))$ is a suspension, there is a homotopy equivalence

$$P^n(p^r) \ltimes (\bigvee_{i=2}^{m} P^{n+1}(p^r)) \simeq P^{2n+1}(p^r) \vee G.$$

In particular, the map a in (7.2) has a right homotopy inverse. The homotopy commutativity of the bottom right square in (7.2) then implies that b also has a right homotopy inverse. Thus

$$\widetilde{C} \simeq S^{2n+1} \vee G.$$

Since G is the homotopy cofibre of ℓ', the left homotopy inverse of $(1 \wedge q_t) \circ q \circ \ell'$ in Lemma 7.4 implies that

$$G \simeq \bigvee_{i=2}^{m} P^{n+1}(p^r) \vee \Big(P^n(p^r) \wedge (\bigvee_{\substack{i=2\\ i\neq t}}^{m} P^{n+1}(p^r)) \Big) \vee P^{2n}(p^r).$$

In particular, the upper right vertical arrow in (7.1) composed with $\widetilde{C} \longrightarrow G$ is the inclusion of $\bigvee_{i=2}^m P^{n+1}(p^r)$ into G. Therefore, the map $\bigvee_{i=2}^m P^{n+1}(p^r) \longrightarrow \widetilde{C}$ in (7.1) has a left homotopy inverse, proving part (a). Further, this implies that its homotopy cofibre $\overline{C}$ satisfies

$$\overline{C} \simeq \Big(P^n(p^r) \wedge (\bigvee_{\substack{i=2\\ i\neq t}}^{m} P^{n+1}(p^r)) \Big) \vee (S^{2n+1} \vee P^{2n}(p^r)).$$

Finally, note that the inclusion of S^{2n+1} in $\overline{C}$ is via the composite $S^{2n+1} \longrightarrow \widetilde{C} \longrightarrow \overline{C}$, completing the proof of part (b). □

Recall that for $1 \leq k \leq m$ the map $i_k \colon P^{n+1}(p^r) \longrightarrow \bigvee_{i=1}^m P^{n+1}(p^r)$ is the inclusion of the kth-wedge summand. The Whitehead product $[i_j, i_k]$ is a map $\Sigma P^n(p^r) \wedge P^n(p^r) \longrightarrow \bigvee_{i=1}^m P^{n+1}(p^r)$. Combining this with the map v in Example 7.2 gives a composite

$$S^{2n} \xrightarrow{\Sigma v} \Sigma P^n(p^r) \wedge P^n(p^r) \xrightarrow{[i_j,i_k]} \bigvee_{i=1}^{m} P^{n+1}(p^r)$$

where Σv induces an inclusion in mod-p homology. Note that as v factors as the composite $S^{2n} \longrightarrow P^{2n+1}(p^r) \longrightarrow \Sigma P^n(p^r) \wedge P^n(p^r)$, where the left map is the inclusion of the bottom cell and the right map is the inclusion of the top dimensional Moore space, the map $[i_j, i_k] \circ v$ can alternatively be regarded as $S^{2n} \longrightarrow P^{2n+1}(p^r) \xrightarrow{[i_j,i_k]_r} \bigvee_{i=1}^k P^{n+1}(p^r)$, where $[i_j, i_k]_r$ is the mod-p^r Whitehead product defined before Lemma 6.11. In what follows the notation will be formulated in terms of $[i_j, i_k] \circ \Sigma v$.

THEOREM 7.6. *Let p be an odd prime, $r \geq 1$ and $n \geq 2$. Suppose that there is a homotopy cofibration*

$$S^{2n} \xrightarrow{f} \bigvee_{i=1}^{m} P^{n+1}(p^r) \longrightarrow M.$$

Suppose also that $f = f_1 + f_2$ where:

- $f_1 = \sum_{j=2}^m [i_1, i_j] \circ (d_j \cdot \Sigma v)$ *for* $d_j \in \mathbb{Z}$;
- *there is at least one* $t \in \{2, \dots, m\}$ *such that the mod-p reduction of d_t is a unit;*
- f_2 *factors as* $S^{2n} \xrightarrow{\gamma} \bigvee_{i=2}^m P^{n+1}(p^r) \xrightarrow{I} \bigvee_{i=1}^m P^{n+1}(p^r)$ *for some map* γ;
- $\Sigma\gamma$ *is null homotopic.*

Then there is a homotopy fibration

$$(\Omega P^{n+1}(p^r) \ltimes \overline{C}) \vee (\bigvee_{i=2}^{m} P^{n+1}(p^r)) \longrightarrow M \xrightarrow{q'} P^{n+1}(p^r)$$

where $\overline{C} \simeq \left(P^n(p^r) \wedge (\bigvee_{\substack{i=2\\ i\neq t}}^{m} P^{n+1}(p^r))\right) \vee \left(S^{2n+1} \vee P^{2n}(p^r)\right)$, *and this homotopy fibration splits after looping to give a homotopy equivalence*

$$\Omega M \simeq \Omega P^{n+1}(p^r) \times \Omega\Big((\Omega P^{n+1}(p^r) \ltimes \overline{C}) \vee (\bigvee_{i=2}^{m} P^{n+1}(p^r))\Big).$$

Proof. The proof proceeds in several steps.

Step 1 (An observation). We are assuming that $f_1 = \sum_{j=2}^{m}[i_1, i_j] \circ (d_j \cdot \Sigma v)$ so the map $\Sigma A \xrightarrow{h_{1,j}} \Sigma X_1 \wedge X_j$ in Proposition 6.4 is $S^{2n} \xrightarrow{d_j \cdot \Sigma v} \Sigma P^n(p^r) \wedge P^n(p^r)$ in this case. This does not have a left homotopy inverse, regardless of the value of d_j. Therefore the hypotheses of Proposition 6.4 do not apply. However, by Lemma 4.20, the map $P^n(p^r) \wedge S^{2n} \xrightarrow{1\wedge d_j \cdot \Sigma v} P^n(p^r) \wedge \Sigma P^n(p^r) \wedge P^n(p^r)$ does have a left homotopy inverse if the mod-p reduction of d_t is a unit. Therefore, by Lemma 7.1, the map

$$\Omega P^{n+1}(p^r) \ltimes S^{2n} \xrightarrow{1\wedge d_j \cdot \Sigma v} \Omega P^{n+1}(p^r) \wedge (\Sigma P^n(p^r) \wedge P^n(p^r))$$

has a left homotopy inverse.

Step 2 (The approach). In light of Step 1, the approach is to modify the proof of Theorem 5.8, on which the proof of Proposition 6.4 relied. Consider the data

$$\begin{array}{ccccc}
 & & E & \longrightarrow & E' \\
 & & \downarrow & & \downarrow \\
S^{2n} & \xrightarrow{f} & \bigvee_{i=1}^{m} P^{n+1}(p^r) & \longrightarrow & M \\
 & & \downarrow{\scriptstyle q_1} & & \downarrow{\scriptstyle q'} \\
 & & P^{n+1}(p^r) & = & P^{n+1}(p^r).
\end{array}$$

The map q' exists since $f = f_1 + f_2$ where f_1 factors through the sum of the Whitehead products $[\iota_i, \iota_j]$ and so composes trivially with q_1, while f_2 factors through I and so composes trivially with q_1. Arguing just as for Steps 1 through 3 in the proof of Proposition 6.4 shows that the hypotheses of the present lemma imply parts (a) through (d) of Theorem 5.8 - which are identical to parts (a) through (d) of Proposition 5.7. Therefore the $k = 1$ case of Proposition 5.7 implies that there is a homotopy cofibration diagram

$$\begin{array}{ccc}
\Sigma(\Omega P^{n+1}(p^r)\ltimes S^{2n}) & \xlongequal{\quad} & \Sigma(\Omega P^{n+1}(p^r)\ltimes S^{2n}) \\
\big\downarrow{\scriptstyle \Sigma\kappa_1+*} & & \big\downarrow{\scriptstyle \Sigma\theta} \\
\Sigma(\Omega P^{n+1}(p^r)\ltimes (P^n(p^r)\wedge\Sigma D))\vee\Sigma^2 D & \xrightarrow{\Sigma\epsilon} & \Sigma E \\
\big\downarrow{\scriptstyle \Sigma\lambda\vee 1} & & \big\downarrow{\scriptstyle \Sigma\eta} \\
\Sigma(\Omega P^{n+1}(p^r)\ltimes \overline{C})\vee\Sigma^2 D & \xrightarrow{\psi} & \Sigma E'
\end{array} \tag{7.3}$$

where $D = \bigvee_{i=2}^{m} P^{n+1}(p^r)$, ϵ is a homotopy equivalence, λ is the map to the homotopy cofibre of κ_1, η is the map to the homotopy cofibre of θ, and ψ is an induced map of cofibres. As ϵ is a homotopy equivalence the five-lemma implies that ψ induces an isomorphism in homology and so is a homotopy equivalence by Whitehead's Theorem.

STEP 3 (The homotopy type of E', setting up). Write $\psi = \psi_1 \perp \psi_2 \perp \psi_3$ where ψ_1, ψ_2 and ψ_3 are the restrictions of ψ to $\Sigma\overline{C}$, $\Sigma(\Omega P^{n+1}(p^r)\wedge\overline{C})$ and $\Sigma^2 D$ respectively. Similarly write $\epsilon = \epsilon_1 \perp \epsilon_2 \perp \epsilon_3$ where ϵ_1, ϵ_2 and ϵ_3 are the restrictions of ϵ to $P^n(p^r)\wedge\Sigma D$, $\Omega P^{n+1}(p^r)\wedge(P^n(p^r)\wedge\Sigma D)$ and $\Sigma^2 D$ respectively. First, observe that the bottom square in (7.3) implies that $\psi_3 = \Sigma\eta\circ\Sigma\epsilon_3$. In particular, if $\psi_3' = \eta\circ\epsilon_3$ then $\psi_3 = \Sigma\psi_3'$. Second, consider the homotopy cofibration $S^{2n} \xrightarrow{q\circ\ell} P^n(p^r)\wedge\Sigma D \xrightarrow{a} \overline{C}$, which defines the map a. This does not split but, by Step 1, the homotopy cofibration $\Omega P^{n+1}(p^r)\wedge S^{2n} \xrightarrow{1\wedge(q\circ\ell)} \Omega P^{n+1}(p^r)\wedge(P^n(p^r)\wedge\Sigma D) \xrightarrow{1\wedge a} \Omega P^{n+1}(p^r)\wedge\overline{C}$ does split. If b is a right homotopy inverse of $1\wedge a$ then, since $\kappa_3 = 1\ltimes(q\circ\ell)$ by (5.2), the bottom square in (7.3) implies that $\psi_2 \simeq \Sigma\eta\circ\Sigma\epsilon_2\circ\Sigma b$. In particular, if $\psi_2' = \eta\circ\epsilon_2\circ b$ then $\psi_2 \simeq \Sigma\psi_2'$. Third, consider the homotopy cofibration diagram (in which columns are homotopy cofibrations)

$$\begin{array}{ccccc}
S^{2n} & \xrightarrow{j} & \Omega P^{2n+1}(p^r)\ltimes S^{2n} & \xlongequal{\quad} & \Omega P^{2n+1}(p^r)\ltimes S^{2n} \\
\big\downarrow{\scriptstyle \ell} & & \big\downarrow{\scriptstyle 1\ltimes\ell} & & \big\downarrow{\scriptstyle \theta} \\
P^n(p^r)\ltimes\Sigma D & \xrightarrow{j} & \Omega P^{n+1}(p^r)\ltimes(P^n(p^r)\ltimes\Sigma D) & \xrightarrow{e} & E \\
\big\downarrow & & \big\downarrow & & \big\downarrow{\scriptstyle \eta} \\
\widetilde{C} & \xrightarrow{j} & \Omega P^{n+1}(p^r)\ltimes\widetilde{C} & \xrightarrow{\xi} & E'
\end{array} \tag{7.4}$$

where ξ is an induced map of cofibres. The splitting $\widetilde{C}\simeq\Sigma D\vee\overline{C}$ in Lemma 7.5 (a) implies that there is a map

$$\alpha\colon \overline{C} \longrightarrow \widetilde{C} \xrightarrow{\xi\circ j} E'.$$

Note that the map

$$\beta\colon \Sigma D \longrightarrow \widetilde{C} \xrightarrow{\xi\circ j} E'$$

is the same as $\eta\circ e\circ j$ restricted to ΣD.

STEP 4 (The homotopy type of E'). We claim that the map

$$\overline{C}\vee\Sigma(\Omega P^{n+1}(p^r)\wedge\overline{C})\vee\Sigma^2 D \xrightarrow{\alpha\perp\psi_2'\perp\psi_3'} E'$$

is a homotopy equivalence. It suffices to show that it induces an isomorphism in homology. To do this, it suffices to show that it induces an isomorphism in mod-q homology for every prime q and in rational homology. In mod-p homology, since ℓ induces an injection, the map $P^n(p^r) \ltimes \Sigma D \longrightarrow \widetilde{C}$ is a surjection. Thus the image of $(\xi \circ j)_*$ is determined by the image of $(\eta \circ e \circ j)_*$. That is, the image of $(\alpha \perp \beta)_*$ is determined by the image of $(\eta \circ e \circ j)_*$. By definition of ϵ, $e \circ j \simeq \epsilon_1 \perp \epsilon_3$. Thus, after suspending, the bottom square in (7.3) implies that the image of $(\Sigma\eta \circ (\Sigma\epsilon_1 \perp \Sigma\epsilon_3))_*$ equals the image of $(\psi_1 \perp \psi_3)_*$. Since $\beta = \psi_3'$, we obtain that the image of $(\Sigma\alpha \perp \Sigma\beta)_*$ equals the image of $(\psi_1 \perp \psi_3)_*$. Hence as $\psi = \psi_1 \perp \psi_2 \perp \psi_3$ is a homotopy equivalence, it induces an isomorphism in mod-p homology, and therefore so does $\Sigma\alpha \perp \psi_2 \perp \psi_3$. Hence, desuspending, $\alpha \perp \psi_2' \perp \psi_3'$ must induce an isomorphism in mod-p homology. Localizing at a prime q for $q \neq p$ or rationally, since $P^n(p^r)$ is contractible, the homotopy cofibration diagram (7.4) reduces to

$$\begin{array}{ccccc} S^{2n} & = & S^{2n} & = & S^{2n} \\ \downarrow & & \downarrow & & \downarrow \\ * & \longrightarrow & * & \longrightarrow & * \\ \downarrow & & \downarrow & & \downarrow \\ S^{2n+1} & = & S^{2n+1} & = & S^{2n+1}. \end{array}$$

In particular, $\widetilde{C} \simeq S^{2n+1}$ and $\xi \circ j$ is a homotopy equivalence. Observe also that α is a homotopy equivalence. Thus $\alpha \perp \psi_2' \perp \psi_3'$ is a homotopy equivalence and so induces an isomorphism in mod-q, or respectively rational, homology. Hence $\alpha \perp \psi_2' \perp \psi_3'$ induces an isomorphism in integral homology, as required.

Step 5 (The homotopy type of $\overline{C}$). As $f_1 = \sum_{j=2}^m [i_1, i_j] \circ (d_j \cdot \Sigma v)$, the map $\Sigma A \xrightarrow{h_{1,j}} \Sigma X_1 \wedge X_j$ in Proposition 6.4 in this case is $S^{2n} \xrightarrow{d_j \cdot \Sigma v} \Sigma P^n(p^r) \wedge P^n(p^r)$. As there is at least one $t \in \{2, \ldots, m\}$ such that the mod-p reduction of d_j is a unit, the map $S^{2n} \xrightarrow{d_t \cdot \Sigma v} \Sigma P^n(p^r) \wedge P^n(p^r)$ induces an inclusion in mod-p homology. Therefore, by Lemma 7.5, there is a homotopy equivalence $\overline{C} \simeq \left(P^n(p^r) \wedge (\bigvee_{\substack{i=2 \\ i \neq t}}^{m} P^{n+1}(p^r))\right) \vee \left(S^{2n+1} \vee P^{2n}(p^r)\right)$.

□

As a special case of Theorem 7.6 we prove Theorem 1.3, restated verbatim.

Theorem 7.7. *Let p be an odd prime, $r \geq 1$ and $n \geq 2$. Suppose that there is a homotopy cofibration*

$$S^{2n} \xrightarrow{f} \bigvee_{i=1}^{m} P^{n+1}(p^r) \longrightarrow M$$

where $f = \sum_{1 \leq j < k \leq m} [i_j, i_k] \circ (d_{j,k} \cdot v)$ for $d_{j,k} \in \mathbb{Z}$ and at least one $d_{j,k}$ reduces to a unit mod-p. Rearranging the wedge summands $\bigvee_{i=1}^m P^{n+1}(p^r)$ so that some $d_{1,t}$ reduces to a unit mod-p, there is a homotopy fibration

$$(\Omega P^{n+1}(p^r) \ltimes \overline{C}) \vee (\bigvee_{i=2}^{m} P^{n+1}(p^r)) \longrightarrow M \xrightarrow{q'} P^{n+1}(p^r)$$

where $\overline{C} \simeq \Big(P^n(p^r) \wedge (\bigvee_{\substack{i=2\\ i\neq t}}^{m} P^{n+1}(p^r))\Big) \vee \Big(S^{2n+1} \vee P^{2n}(p^r)\Big)$, *and this homotopy fibration splits after looping to give a homotopy equivalence*

$$\Omega M \simeq \Omega P^{n+1}(p^r) \times \Omega\Big((\Omega P^{n+1}(p^r) \ltimes \overline{C}) \vee (\bigvee_{i=2}^{m} P^{n+1}(p^r))\Big).$$

PROOF. By hypothesis, there is a homotopy cofibration

$$S^{2n} \xrightarrow{f} \bigvee_{i=1}^{m} P^{n+1}(p^r) \longrightarrow M$$

where $f = \sum_{1\leq j<k\leq m}[i_j, i_k] \circ (d_{j,k} \cdot v)$ for $d_{j,k} \in \mathbb{Z}$ and at least one $d_{j,k}$ reduces to a unit mod-p. Rearrange the wedge summands $\bigvee_{i=1}^{m} P^{n+1}(p^r)$ so that at least one $d_{1,t}$ reduces to a unit mod-p. Let $f_1 = \sum_{k=2}^{m}[i_1, i_k] \circ .(d_{1,k} \cdot v)$ and $f_2 = \sum_{2\leq j<k\leq m}[i_j, i_k] \circ (d_{j,k} \cdot v)$. Then $f = f_1 + f_2$, there is a $t \in \{2, \ldots, m\}$ such that the mod-p reduction of $d_{1,t}$ is a unit, as f_2 does not involve i_1 it factors as a composite $S^{2n} \xrightarrow{\gamma} \bigvee_{i=2}^{m} P^{n+1}(p^r) \xrightarrow{I} \bigvee_{i=1}^{m} P^{n+1}(p^r)$ where γ is the same sum of Whitehead products as in f_2 but with each i_j for $2 \leq j \leq k$ throught of as having range $\bigvee_{i=2}^{m} P^{n+1}(p^r)$ rather than $\bigvee_{i=1}^{m} P^{n=1}(p^r)$, and $\Sigma\gamma$ is null homotopic since it is a sum of Whitehead products. Thus the attaching map f satisfies the hypotheses of Theorem 7.6, implying that there is a homotopy fibration

$$(\Omega P^{n+1}(p^r) \ltimes \overline{C}) \vee (\bigvee_{i=2}^{m} P^{n+1}(p^r)) \longrightarrow M \xrightarrow{q'} P^{n+1}(p^r)$$

where $\overline{C} \simeq \Big(P^n(p^r) \wedge (\bigvee_{\substack{i=2\\ i\neq t}}^{m} P^{n+1}(p^r))\Big) \vee \Big(S^{2n+1} \vee P^{2n}(p^r)\Big)$, and this homotopy fibration splits after looping to give a homotopy equivalence

$$\Omega M \simeq \Omega P^{n+1}(p^r) \times \Omega\Big((\Omega P^{n+1}(p^r) \ltimes \overline{C}) \vee (\bigvee_{i=2}^{m} P^{n+1}(p^r))\Big).$$

□

EXAMPLE 7.8. In Theorem 7.6, or Theorem 1.3, if the cofibration takes the form

$$S^{2n} \xrightarrow{f} \bigvee_{i=1}^{2m} P^{n+1}(p^r) \longrightarrow M$$

where $f = [i_1, i_2] \circ v + [i_3, i_4] \circ v + \cdots + [i_{2m-1}, i_{2m}] \circ v$ then M is an $(n-1)$-connected $(2n+1)$-dimensional Poincaré Duality complex. In fact, it is the m-fold connected sum $N\#\cdots\#N$ of the Poincaré Duality complex N defined by the homotopy cofibration

$$S^{2n} \xrightarrow{[i_1,i_2]\circ v} P^{n+1}(p^r) \vee P^{n+1}(p^r) \longrightarrow N.$$

Theorem 7.6 then gives a homotopy decomposition of ΩM.

EXAMPLE 7.9. Modifying the previous example, consider a homotopy cofibration

$$S^{2n} \xrightarrow{f'} \bigvee_{i=1}^{2m} P^{n+1}(p^r) \longrightarrow M'$$

where $f' = [\imath_1, \imath_2] \circ v + [\imath_3, \imath_4] \circ v + \cdots + [\imath_{2m-1}, \imath_{2m}] \circ v + f''$. Here, f'' is a composite $f''\colon S^{2n} \xrightarrow{\gamma} \bigvee_{i=2}^{2m} P^{n+1}(p^r) \xrightarrow{I} \bigvee_{i=1}^{2m} P^{n+1}(p^r)$ with the property that $\Sigma\gamma$ is null homotopic. Possibly γ is a sum of more Whitehead products, possibly it is a class of finite order, or some combination of the two. Then M' may or may not be a Poincaré Duality complex but Theorem 7.6 still applies, giving a homotopy decomposition of $\Omega M'$. Note that the decompositions for $\Omega M'$ and the space ΩM in the previous example are identical. That is, while f'' may mean $M \not\simeq M'$, after looping we nevertheless have $\Omega M \simeq \Omega M'$.

CHAPTER 8

Inert maps

Recall from the Introduction that if $\Sigma A \xrightarrow{f} Y \xrightarrow{h} Y'$ is a homotopy cofibration then the map f is *inert* if Ωh has a right homotopy inverse. An interesting example we have already seen is the homotopy cofibration $X^{\wedge k} \wedge \Sigma Y \xrightarrow{ad^k(i_1)(i_2)} \Sigma X \vee \Sigma Y \xrightarrow{m_k} M_k$ in Chapter 4. By Lemma 4.12, Ωm_k has a right homotopy inverse, and hence $ad^k(i_1)(i_2)$ is inert.

The inert property is exactly one of the main hypotheses of Theorem 3.5, and that theorem will play a key role in what follows. As such, it is useful to recall what it says, compressed slightly to only what will be needed subsequently. Suppose that $\Sigma A \xrightarrow{f} Y \xrightarrow{h} Y'$ is a homotopy cofibration and E is the homotopy fibre of h. If Ωh has a right homotopy inverse then there is a homotopy equivalence

$$\Omega Y' \ltimes \Sigma A \xrightarrow{\theta} E$$

and a homotopy fibration

$$(\Omega Y' \wedge \Sigma A) \vee \Sigma A \xrightarrow{[\gamma, f]+f} Y \xrightarrow{\quad h \quad} Y'$$

where γ is the composite $\Sigma\Omega Y' \xrightarrow{\Sigma s} \Sigma\Omega Y \xrightarrow{ev} Y$.

Suppose that there are homotopy cofibrations

$$\Sigma A \xrightarrow{f} X \xrightarrow{j} M$$

and

$$\Sigma A \xrightarrow{g} Y \xrightarrow{k} N.$$

Since ΣA is a suspension, f and g can be added: $f+g$ is the composite

$$f+g\colon \Sigma A \xrightarrow{\sigma} \Sigma A \vee \Sigma A \xrightarrow{f \vee g} X \vee Y$$

where σ is the comultiplication on ΣA. Define C by the homotopy cofibration

$$\Sigma A \xrightarrow{f+g} X \vee Y \longrightarrow C.$$

Let $q_1 : X \vee Y \longrightarrow X$ be the pinch map to the first wedge summand. Then there is a homotopy cofibration diagram

$$\begin{array}{ccccc} \Sigma A & \xrightarrow{f+g} & X \vee Y & \longrightarrow & C \\ \| & & \downarrow{\scriptstyle q_1} & & \downarrow{\scriptstyle \varphi} \\ \Sigma A & \xrightarrow{\quad f \quad} & X & \xrightarrow{\quad j \quad} & M \end{array} \tag{8.1}$$

that defines the map φ. Let h be the composite

$$h\colon X \vee Y \xrightarrow{q_1} X \xrightarrow{j} M.$$

Note that by (8.1), h is homotopic to the composite $X \vee Y \longrightarrow C \xrightarrow{\varphi} M$. Let E and E' be the homotopy fibres of h and φ respectively. Then we obtain the following diagram of spaces and maps that collects the data that will go into Theorem 2.2:

$$(8.2) \qquad \begin{array}{ccccc} & & E & \longrightarrow & E' \\ & & \downarrow & & \downarrow \\ \Sigma A & \xrightarrow{f+g} & X \vee Y & \longrightarrow & C \\ & & \downarrow{\scriptstyle h} & & \downarrow{\scriptstyle \varphi} \\ & & M & = & M. \end{array}$$

LEMMA 8.1. *Suppose that f is inert. Then both Ωh and $\Omega\varphi$ have right homotopy inverses.*

PROOF. As f is inert there is a map $t\colon \Omega M \longrightarrow \Omega X$ such that $\Omega j \circ t$ is homotopic to the identity map on ΩM. Consider the composite

$$\Omega M \xrightarrow{t} \Omega X \xrightarrow{\Omega i_1} \Omega(X \vee Y) \xrightarrow{\Omega h} \Omega M$$

where i_1 is the inclusion of the left wedge summand. By definition, $h = j \circ q_1$, so as $q_1 \circ i_1$ is the identity map on X, we obtain

$$\Omega h \circ \Omega i_1 \circ t \simeq \Omega j \circ \Omega q_1 \circ \Omega i_1 \circ t \simeq \Omega j \circ t \simeq id_{\Omega M}.$$

Thus Ωh has a right homotopy inverse. The homotopy commutativity of the bottom square in (8.2) then implies that $\Omega\varphi$ also has a right homotopy inverse. □

Since Ωh has a right homotopy inverse, applying Theorem 2.2 to (8.2) gives a homotopy cofibration

$$(8.3) \qquad \Omega M \ltimes \Sigma A \xrightarrow{\theta_{f+g}} E \longrightarrow E'.$$

Next consider the homotopy cofibration $\Sigma A \xrightarrow{f} X \xrightarrow{j} M$. Let F be the homotopy fibre of j. Since f is inert the map Ωj has a right homotopy inverse so by Theorem 3.5 there is a homotopy equivalence

$$(8.4) \qquad \Omega M \ltimes \Sigma A \xrightarrow{\theta_f} F.$$

The homotopy cofibrations (8.3) and (8.4) can be put together. By definition, $h = j \circ q_1$, so there is a homotopy fibration diagram

$$(8.5) \qquad \begin{array}{ccc} E & \xrightarrow{\ell} & F \\ \downarrow & & \downarrow \\ X \vee Y & \xrightarrow{q_1} & X \\ \downarrow{\scriptstyle h} & & \downarrow{\scriptstyle j} \\ M & = & M. \end{array}$$

where ℓ is the induced map of fibres. The right homotopy inverse $s\colon \Omega M \longrightarrow \Omega(X \vee Y)$ for Ωh implies that $\Omega q_1 \circ s$ is a right homotopy inverse for Ωj. The naturality property in Remark 2.7 implies that there is a homotopy commutative diagram of cofibrations

$$\begin{array}{ccccc} \Omega M \ltimes \Sigma A & \xrightarrow{\theta_{f+g}} & E & \longrightarrow & E' \\ \| & & \downarrow{\scriptstyle \ell} & & \downarrow \\ \Omega M \ltimes \Sigma A & \xrightarrow{\theta_f} & F & \longrightarrow & * \end{array} \tag{8.6}$$

Note that θ_f being a homotopy equivalence implies that the map θ_{f+g} has a left homotopy inverse. Moreover, this inverse is independent of g. We record this for future reference.

LEMMA 8.2. *The map $\Omega M \ltimes \Sigma A \xrightarrow{\theta_{f+g}} E$ has a left homotopy inverse that is independent of the map g.* □

Next, consider the special case when g is the trivial map. In (8.2) the homotopy cofibration $\Sigma A \xrightarrow{f+g} X \vee Y \longrightarrow C$ becomes $\Sigma A \xrightarrow{f+*} X \vee Y \xrightarrow{j\vee 1} M \vee Y$ and the map $C \xrightarrow{\varphi} M$ can be chosen to be the pinch map $M \vee Y \xrightarrow{q_1} M$. Therefore the homotopy fibre E' of φ becomes $\Omega M \ltimes Y$. Hence the homotopy cofibration (8.3) becomes

$$\Omega M \ltimes \Sigma A \xrightarrow{\theta_{f+*}} E \longrightarrow \Omega M \ltimes Y. \tag{8.7}$$

In this case we show that the cofibration (8.7) splits in a way that behaves well with respect to the map ℓ in (8.6).

LEMMA 8.3. *The map $E \longrightarrow \Omega M \ltimes Y$ in (8.7) has a right homotopy inverse $r\colon \Omega M \ltimes Y \longrightarrow E$ such that $\ell \circ r$ is null homotopic.*

PROOF. The identifications in (8.2) when $g = *$ imply that there is a homotopy fibration diagram

$$\begin{array}{ccc} E & \longrightarrow & \Omega M \ltimes Y \\ \downarrow & & \downarrow \\ X \vee Y & \xrightarrow{j\vee 1} & M \vee Y \\ \downarrow{\scriptstyle h} & & \downarrow{\scriptstyle q_1} \\ M & = & M. \end{array}$$

In particular, the upper square is a homotopy pullback. From the naturality of the pinch map q_1 we obtain a pullback map

$$\begin{array}{ccccc} \Omega X \ltimes Y & & & & \\ & \overset{\overline{r}}{\searrow} & & \overset{\Omega j \ltimes 1}{\searrow} & \\ & & E & \longrightarrow & \Omega M \ltimes Y \\ & & \downarrow & & \downarrow \\ & & X \vee Y & \xrightarrow{j\vee 1} & M \vee Y \end{array}$$

that defines $\overline{r}$. Since f is inert, Ωj has a right homotopy inverse $t\colon \Omega M \longrightarrow \Omega X$. Let r be the composite

$$r\colon \Omega M \ltimes Y \xrightarrow{t\ltimes 1} \Omega X \ltimes Y \xrightarrow{\overline{r}} E.$$

Then the previous diagram implies that the composite $\Omega M \ltimes Y \xrightarrow{r} E \longrightarrow \Omega M \ltimes Y$ is homotopic to $(\Omega j \ltimes 1) \circ (t \ltimes 1)$, which is homotopic to the identity map. Thus $E \longrightarrow \Omega M \ltimes Y$ has a right homotopy inverse.

It remains to show that $\ell \circ r$ is null homotopic. Consider the diagram

$$\begin{array}{ccccccc} \Omega M \ltimes Y & \xrightarrow{t\ltimes 1} & \Omega X \ltimes Y & \xrightarrow{\overline{r}} & E & \xrightarrow{\ell} & F \\ &&&& \big\downarrow && \big\downarrow \\ &&&& X \vee Y & \xrightarrow{q_1} & X. \end{array}$$

The square homotopy commutes by (8.5). The definition of $\overline{r}$ as a pullback map implies that the composite $\Omega X \ltimes Y \xrightarrow{\overline{r}} E \longrightarrow X \vee Y$ is the map from the homotopy fibre of q_1 to the total space. Therefore composing it with q_1 is null homotopic so the lower direction around the diagram is null homotopic. Hence the upper direction around the diagram is null homotopic. By definition, $r = \overline{r} \circ (t \ltimes 1)$, implying that $\ell \circ r$ is null homotopic when composed with $F \longrightarrow X$. Now consider the homotopy fibration sequence $\Omega M \xrightarrow{\partial} F \longrightarrow X \xrightarrow{j} M$, where ∂ is the connecting map. On the one hand, we have just seen that $\ell \circ r$ must lift through ∂. On the other hand, since Ωj has a right homotopy inverse, ∂ is null homotopic. Therefore $\ell \circ r$ is null homotopic, as asserted. $\square$

In general, suppose that $U \xrightarrow{s} V \xrightarrow{t} W$ is a homotopy cofibration where t has a right homotopy inverse r. Then the composite $e \colon U \vee W \xrightarrow{s \vee r} V \vee V \xrightarrow{\nabla} V$ induces an isomorphism in homology, where ∇ is the fold map. Thus if U, V and W are simply-connected then e is a homotopy equivalence by Whitehead's Theorem. In our case, we are assuming that all spaces are simply-connected, so the existence of the right homotopy inverse in Lemma 8.3 implies the following.

COROLLARY 8.4. *From the homotopy cofibration* $\Omega M \ltimes \Sigma A \xrightarrow{\theta_{f+*}} E \longrightarrow \Omega M \ltimes Y$ *we obtain a homotopy equivalence*

$$(\Omega M \ltimes \Sigma A) \vee (\Omega M \ltimes Y) \xrightarrow{\theta_{f+*} \vee r} E \vee E \xrightarrow{\quad\nabla\quad} E$$

where ∇ *is the fold map.* $\square$

Now return to the general case of the homotopy cofibration $\Omega M \ltimes \Sigma A \xrightarrow{\theta_{f+g}} E \longrightarrow E'$. We will use the special case when $g = *$ to show a splitting in the general case, and identify the homotopy type of E'. This requires a preliminary lemma, which is stated abstractly. To distinguish identity maps on different spaces, for a space V let $1_V \colon V \longrightarrow V$ be the identity map on V.

LEMMA 8.5. *Suppose that there are homotopy cofibrations* $P \xrightarrow{p} Q \xrightarrow{j_p} R_p$ *and* $P \xrightarrow{q} Q \xrightarrow{j_q} R_q$ *where all spaces are simply-connected. Also suppose that there are maps* $Q \xrightarrow{k} P$ *and* $R_q \xrightarrow{s} Q$ *such that* $k \circ p \simeq 1_P$, $k \circ q \simeq 1_P$, $j_q \circ s \simeq 1_{R_q}$ *and* $k \circ s \simeq *$. *Then the composite* $R_q \xrightarrow{s} Q \xrightarrow{j_p} R_p$ *is a homotopy equivalence.*

PROOF. Start with the homotopy cofibration $P \xrightarrow{q} Q \xrightarrow{j_q} R_q$. Since all spaces are simply-connected, the fact that $j_q \circ s \simeq 1_{R_q}$ implies that the composite

$$e \colon P \vee R_q \xrightarrow{p \vee s} Q \vee Q \xrightarrow{\nabla} Q$$

is a homotopy equivalence, where ∇ is the fold map. Since $k \circ q \simeq 1_P$, $k \circ s \simeq *$, and the fold map is natural, we obtain a homotopy commutative square

$$\begin{array}{ccc} P \vee R_q & \xrightarrow{e} & Q \\ \downarrow{\scriptstyle q_1} & & \downarrow{\scriptstyle k} \\ P & = & P \end{array}$$

where q_1 is the pinch map to the first wedge summand. Restricting to R_q we therefore obtain a homotopy cofibration

$$R_q \xrightarrow{s} Q \xrightarrow{k} P.$$

Now the fact that $k \circ p \simeq 1_P$ implies that we obtain a homotopy pushout diagram

$$\begin{array}{ccccc} & & R_q & = & R_q \\ & & \downarrow{\scriptstyle s} & & \downarrow \\ P & \xrightarrow{p} & Q & \xrightarrow{j_p} & R_p \\ \| & & \downarrow{\scriptstyle k} & & \downarrow \\ P & \xrightarrow{=} & P & \longrightarrow & *. \end{array}$$

To be clear, $k \circ p \simeq 1_P$ implies that the homotopy cofibre along the bottom row is trivial, and therefore the homotopy cofibre of $j_p \circ s$ is trivial. Hence $j_p \circ s$ induces an isomorphism in homology and so, as spaces are simply-connected, it is a homotopy equivalence by Whitehead's Theorem. □

THEOREM 8.6. *Suppose that there are homotopy cofibrations* $\Sigma A \xrightarrow{f} X \longrightarrow M$ *and* $\Sigma A \xrightarrow{g} Y \longrightarrow N$ *where f is inert. Define the homotopy cofibration* $\Sigma A \xrightarrow{f+g} X \vee Y \longrightarrow C$ *and the homotopy fibration* $E' \longrightarrow C \xrightarrow{\varphi} M$ *as in* (8.2). *Then the following hold:*

(a) *the composite* $\Omega M \ltimes Y \xrightarrow{r} E \longrightarrow E'$ *is a homotopy equivalence, implying that there is a homotopy fibration* $\Omega M \ltimes Y \longrightarrow C \xrightarrow{\varphi} M$;
(b) *there is a homotopy equivalence* $\Omega C \simeq \Omega M \times \Omega(\Omega M \ltimes Y)$;
(c) $f + g$ *is inert, that is, the map* $\Omega(X \vee Y) \longrightarrow \Omega(M \# N)$ *has a right homotopy inverse;*
(d) *there is a homotopy fibration*

$$(\Omega C \wedge \Sigma A) \vee \Sigma A \xrightarrow{\Psi} X \vee Y \longrightarrow C$$

where $\Psi = [\gamma, f + g] + (f + g)$.

PROOF. By (8.3) and (8.7) there are homotopy cofibrations

$$\Omega M \ltimes \Sigma A \xrightarrow{\theta_{f+g}} E \longrightarrow E'$$

and

$$\Omega M \ltimes \Sigma A \xrightarrow{\theta_{f+*}} E \longrightarrow \Omega M \ltimes Y.$$

By Lemma 8.2 there is a map $t\colon E \longrightarrow \Omega M \ltimes \Sigma A$ such that $t \circ \vartheta_{f+g}$ and $t \circ \vartheta_{f+*}$ are both homotopic to the identity map on $\Omega M \ltimes \Sigma A$. By Lemma 8.3 there is a map $\Omega M \ltimes Y \xrightarrow{r} E$ such that the composite $\Omega M \ltimes Y \xrightarrow{r} E \longrightarrow \Omega M \ltimes Y$ is homotopic to the identity map and $t \circ r$ is null homotopic. Therefore, by Lemma 8.5 the composite $\Omega M \ltimes Y \xrightarrow{r} E \longrightarrow E'$ is a homotopy equivalence. This proves part (a).

For part (b), consider the homotopy fibration $E' \longrightarrow C \xrightarrow{\varphi} M$. By Lemma 8.1, $\Omega\varphi$ has a right homotopy inverse. This immediately implies that there is a homotopy equivalence $\Omega C \simeq \Omega M \times \Omega E'$. Now substitute in the homotopy equivalence for E' in part (a) to obtain the asserted homotopy equivalence.

For part (c), let $i \colon X \vee Y \longrightarrow C$ denote the map to the cofibre of $f+g$. To say that $f+g$ is inert means that Ωi has a right homotopy inverse. To see this is the case, consider the loops on the homotopy pullback diagram in (8.2),

$$(8.8) \qquad \begin{array}{ccc} \Omega E & \longrightarrow & \Omega E' \\ \downarrow & & \downarrow \\ \Omega(X \vee Y) & \xrightarrow{\Omega i} & \Omega C \\ \downarrow{\scriptstyle \Omega h} & & \downarrow{\scriptstyle \Omega\varphi} \\ \Omega M & = & \Omega M. \end{array}$$

We check that the homotopy equivalence in part (b) can be chosen to factor through Ωi. First, by Lemma 8.1, the map Ωh has a right homotopy inverse $s \colon \Omega M \longrightarrow \Omega(X \vee Y)$. Thus $\Omega i \circ s$ is a right homotopy inverse for $\Omega\varphi$. Second, by part (a) the composite $\Omega M \ltimes Y \xrightarrow{r} E \longrightarrow E'$ is a homotopy equivalence. Let r' be the composite $\Omega M \ltimes Y \xrightarrow{r} E \longrightarrow X \vee Y$. Then the homotopy commutativity of (8.8) and the fact that Ωi is an H-map implies that the composite

$$\Omega M \times \Omega(\Omega M \ltimes Y) \xrightarrow{s \times \Omega r'} \Omega(X \vee Y) \times \Omega(X \vee Y) \xrightarrow{\mu} \Omega(X \vee Y) \xrightarrow{\Omega i} \Omega C$$

is a homotopy equivalence, where μ is the loop multiplication.

Finally, now knowing that $f+g$ is inert by (c), part (d) is an immediate consequence of Theorem 3.5 applied to the homotopy cofibration $A \xrightarrow{f+g} X \vee Y \longrightarrow C$. □

Remark 8.7. Theorem 8.6 says something notable. The fact that f is inert implies that $f+g$ is inert, regardless of what g is.

CHAPTER 9

Based loops on connected sums

In this section we apply Theorem 8.6 to analyze the based loops on a connected sum of simply-connected Poincaré Duality complexes and prove Theorem 1.4. Suppose that M and N are simply-connected Poincaré Duality complexes of dimension n, where $n \geq 3$. Let X and Y be the $(n-1)$-skeletons of M and N respectively. Then there are homotopy cofibrations

$$S^{n-1} \xrightarrow{f} X \longrightarrow M$$
$$S^{n-1} \xrightarrow{g} Y \longrightarrow N$$

where f and g are the attaching maps for the top cells of M and N respectively. The connected sum $M\#N$ is given by the homotopy cofibration

$$S^{n-1} \xrightarrow{f+g} X \vee Y \longrightarrow M\#N.$$

This is exactly the situation considered in the previous section, taking $A = S^{n-2}$ and $C = M\#N$. Note that as $n \geq 3$ the space S^{n-1} is a simply-connected suspension. As in Chapter 8, there is a map $M\#N \xrightarrow{\varphi} M$, where explicitly in this case it is the map given by collapsing $Y \subseteq M\#N$ to a point. So from Theorem 8.6 we immediately obtain the following, which is a more comprehensive version of Theorem 1.4.

THEOREM 9.1. *Let M and N be simply-connected Poincaré Duality complexes of dimension n, where $n \geq 3$. If the attaching map f for the top cell of M is inert then the following hold:*

(a) *there is a homotopy fibration $\Omega M \ltimes Y \longrightarrow M\#N \xrightarrow{\varphi} M$;*
(b) *there is a homotopy equivalence $\Omega(M\#N) \simeq \Omega M \times \Omega(\Omega M \ltimes Y)$;*
(c) *the attaching map $f+g$ for the top cell of $M\#N$ is inert, that is, the loop map $\Omega(X \vee Y) \longrightarrow \Omega(M\#N)$ has a right homotopy inverse;*
(d) *there is a homotopy fibration*

$$(\Sigma\Omega(M\#N) \wedge S^{n-1}) \vee S^{n-1} \xrightarrow{\Psi} X \vee Y \longrightarrow M\#N$$

where $\Psi = [\gamma, f+g] + (f+g)$. □

We now give several examples of Theorem 9.1. First, we consider taking the connected sum with an $(n-1)$-connected $2n$-dimensional Poincaré Duality complex.

PROPOSITION 9.2. *Let M be an $(n-1)$-connected $2n$-dimensional Poincaré Duality complex such that $n \geq 2$ and the attaching map for the top cell of M is inert. Let N be an $(n-k)$-connected, $2n$-dimensional Poincaré Duality complex with $n-k \geq 1$ and $3k-2 \leq n$. Let $Y = N - *$. Then Y is a suspension and there is a homotopy equivalence*

$$\Omega(M\#N) \simeq \Omega M \times \Omega((\Omega M \wedge Y) \vee Y).$$

PROOF. It is well known that if $m \geq 2$ and V is an $(m-1)$-connected CW-complex of dimension at most $2m-1$ then V is homotopy equivalent to a suspension. In our case, Y is $(m-1)$-connected for $m = n-k+1$, the condition $n-k \geq 1$ implies Y is simply-connected, and the condition that $3k-2 \leq n$ implies that Y is of dimension $\leq 2m-1$. Therefore Y is a suspension.

Since the attaching map for the top cell of M is inert, by Theroem 9.1 (b), there is a homotopy equivalence $\Omega(M\#N) \simeq \Omega M \times \Omega(\Omega M \ltimes Y)$. Since Y is a suspension, there is a homotopy equivalence $\Omega M \ltimes Y \simeq (\Omega M \wedge Y) \vee Y$, and the assertion follows. □

The hypotheses of Proposition 9.2 hold in a wide variety of cases. By [**BT2**], if $n \notin \{4,8\}$ then the attaching map for the top cell of an $(n-1)$-connected $2n$-dimensional Poincaré Duality complex M is inert. If $n \in \{4,8\}$ then the attaching map for the top cell is not known to be inert in all cases but it may be inert for specific cases: for example, the attaching maps for the top cells in $S^4 \times S^4$ and $S^8 \times S^8$ are both inert.

Observe that if $n = 2$ or $n = 3$ then the condition $3k-2 \leq n$ implies $k = 1$, and N is then either a simply-connected four-manifold if $n = 2$ or a 2-connected 6-manifold if $n = 3$; both cases are then simply repeating known decompositions from [**BT1**] or [**BB**]. However, if $n = 4$ then $k = 2$ is valid, so we obtain a homotopy decomposition for $\Omega(M\#N)$ when $M = S^4 \times S^4$ and N is any 2-connected 8-manifold. This is new—in [**BT1**] it was shown that if $H^*(N;\mathbb{Z})$ is torsion-free then such a decomposition exists but Proposition 9.2 dispenses with the torsion-free cohomology condition. In fact, in [**BT1**] it was shown that if $M = S^m \times S^{2n-m}$ and $H^*(N;\mathbb{Z})$ is torsion-free then $\Omega(M\#N)$ decomposes. Proposition 9.2 significantly generalizes this to M being any $(n-1)$-connected $2n$-dimensional Poincaré Duality complex with $n \geq 2$, and N not having a torsion-free cohomology condition but some control over the dimensional range in which the middle cells appear.

Next, we prove a general result in Proposition 9.5 about taking the connected sum with a product and then increasingly specialize it.

LEMMA 9.3. *Let $X_1, \ldots, X_k$ be simply-connected spaces and let $j\colon \bigvee_{i=1}^k X_i \longrightarrow \prod_{i=1}^k X_i$ be the inclusion of the wedge into the product. Then Ωj has a right homotopy inverse.*

PROOF. This is well known. Let $j_i\colon X_i \longrightarrow \bigvee_{i=1}^k X_i$ be the inclusion. Then $j \circ j_i$ is the inclusion of the ith factor in $\prod_{i=1}^k X_i$. Looping to multiply, the product of the maps Ωj_i for $1 \leq i \leq k$ is a right homotopy inverse for Ωj. □

The next lemma gives one source of Poincaré Duality complexes for which the right homotopy inverse hypothesis of Theorem 9.1 holds.

LEMMA 9.4. *Let $k \geq 2$ and suppose that $M_1, \ldots, M_k$ are nontrivial simply-connected finite dimensional Poincaré Duality complexes. Let $M = \prod_{i=1}^k M_i$ and let $J\colon M - * \longrightarrow M$ be the inclusion. Then ΩJ has a right homotopy inverse.*

PROOF. As each M_i is simply-connected, it may be approximated by a CW-complex. Doing so, let d_i be the dimension of M_i. Then $D = \sum_{i=1}^k d_i$ is the dimension of M. As M_i is nontrivial, we have $d_i \geq 1$. Therefore as $k \geq 2$ we obtain $d_i < D$. Thus the inclusion $M_i \longrightarrow M$ factors through the $(D-1)$-skeleton of M, which is homotopy equivalent to $M - *$. Hence the inclusion $\bigvee_{i=1}^k M_i \xrightarrow{j}$

$\prod_{i=1}^{k} M_i = M$ of the wedge into the product factors as a composite $\bigvee_{i=1}^{k} M_i \longrightarrow M - * \xrightarrow{J} M$. By Lemma 9.3, Ωj has a right homotopy inverse. Therefore, so does ΩJ. □

Theorem 9.1 and Lemma 9.4 immediately imply the following.

PROPOSITION 9.5. *Suppose that $M = \prod_{i=1}^{k} M_i$ for $k \geq 2$ and each M_i is a nontrivial simply-connected Poincaré Duality complex of dimension n. Let N be any other simply-connected Poincaré Duality complex of dimension n and let $Y = N - *$. Then there is a homotopy equivalence*

$$\Omega(M\#N) \simeq \Omega M \times \Omega(\Omega M \ltimes Y).$$

We consider special cases of Proposition 9.5 in which the decomposition of $\Omega(M\#N)$ can be further refined.

EXAMPLE 9.6. Suppose that the product M in Proposition 9.5 has dimension $2n$ for $n \geq 2$. Let N be an $(n-1)$-connected $2n$-dimensional Poincaré Duality complex. Then Poincaré Duality implies that $Y = N - *$ is homotopy equivalent to a wedge of d copies of S^n, where d is the rank of $H^*(N;\mathbb{Z})$. If $d \geq 1$ then Y is a suspension, so $\Omega M \ltimes Y \simeq (\Omega M \wedge Y) \vee Y$. Siimilarly, if M is $(2n+1)$-diimensional for $n \geq 2$ and N is an $(n-1)$-connected $(2n+1)$-dimensional Poincaré Duality complex then $Y = N - *$ is homotopy equivalent to a wedge of some number of copies of S^n, S^{n+1} and Moore spaces $P^{n+1}(m)$ for various values of m. Again, if Y is nontrivial then it is a suspension. Therefore, in both cases we obtain a homotopy equivalence

$$\Omega(M\#N) \simeq \Omega M \times ((\Omega M \wedge Y) \vee Y).$$

EXAMPLE 9.7. Suppose that $M = \prod_{i=1}^{k} S^{m_i}$ for $k \geq 2$, each sphere is simply-connected, and N is as in Example 9.6. Since $\Omega M \simeq \prod_{i=1}^{k} \Omega S^{n_i}$, iterating the fact that $\Sigma(X \times Y) \simeq \Sigma X \vee \Sigma Y \vee (\Sigma X \wedge Y)$ and iterating the fact from [**J1**] that

$$\Sigma\Omega S^{m+1} \simeq \bigvee_{r=1}^{\infty} S^{rm+1}$$

shows that $\Sigma\Omega M$ is homotopy equivalent to a wedge of spheres. If M has dimension $2n$ and N is an $(n-1)$-connected $2n$-dimensional Poincaré Duality complex with $Y = N - *$ nontrivial, then Y is homotopy equivalent to a wedge of copies of S^n, implying that $(\Omega M \wedge Y) \vee Y$ is homotopy equivalent to a wedge W of spheres. Thus $\Omega(M\#N) \simeq \Omega M \times \Omega W$. If M has dimension $2n+1$ and N is an $(n-1)$-connected $(2n+1)$-dimensional Poincaré Duality complex with $Y = N - *$ nontrivial, then Y is homotopy equivalent to a wedge of spheres and Moore spaces, implying that $(\Omega M \wedge Y) \vee Y$ is also homotopy equivalent to a wedge W' of spheres and Moore spaces. Thus $\Omega(M\#N) \simeq \Omega M \times \Omega W'$.

EXAMPLE 9.8. Suppose that $M = \prod_{i=1}^{k} \mathbb{C}P^{m_i}$ for $k \geq 2$, M has dimension $2n$, and N is an $(n-1)$-connected $2n$-dimensional Poincaré Duality complex. Since $\Omega\mathbb{C}P^r \simeq S^1 \times \Omega S^{2r+1}$, arguing as in the Example 9.7 shows that $\Sigma\Omega\mathbb{C}P^r$ is homotopy equivalent to a wedge of spheres, as is $\Sigma\Omega M$. Therefore, as in Example 9.7, we obtain a homotopy decomposition of $\Omega(M\#N)$ in terms of ΩM and the loops on a wedge of spheres.

EXAMPLE 9.9. In Example 9.7, suppose that $M = S^{m_1} \times S^{m_2}$, where $m_1+m_2 = 2n$, and N is an $(n-1)$-connected $2n$-dimensional Poincaré Duality complex. The decomposition $\Omega(M\#N) \simeq \Omega M \times \Omega W$ in Example 9.7 therefore implies that $\Sigma\Omega(M\#N) \wedge S^{2n-1}$ is homotopy equivalent to a wedge U of spheres. Theorem 9.1 (d) then implies that there is a homotopy fibration

$$U \vee S^{2n-1} \xrightarrow{\Psi} (S^{m_1} \vee S^{m_2}) \vee Y \longrightarrow (S^{m_1} \times S^{m_2})\#N$$

where the restriction of Ψ to U is a Whitehead product and the restriction of Ψ to S^{2n-1} is the attaching map for the top cell of $(S^{m_1} \times S^{m_2})\#N$. Similarly, if $m_1 + m_2 = 2n+1$ and N is an $(n-1)$-connected $(2n+1)$-dimensional Poincaré Duality complex then the decomposition $\Omega(M\#N) \simeq \Omega M \times \Omega W'$ in Example 9.7 implies that $\Sigma\Omega(M\#N) \wedge S^{2n}$ is homotopy equivalent to a wedge U' of spheres and Moore spaces. Theorem 9.1 (d) then implies that there is a homotopy fibration

$$U' \vee S^{2n} \xrightarrow{\Psi} (S^{m_1} \vee S^{m_2}) \vee Y \longrightarrow (S^{m_1} \times S^{m_2})\#N$$

where the restriction of Ψ to U' is a Whitehead product and the restriction of Ψ to S^{2n} is the attaching map for the top cell of $(S^{m_1} \times S^{m_2})\#N$.

EXAMPLE 9.10. We finish with an interesting specific example. Let X be the Wu manifold, which is a 1-connected 5-manifold whose mod-2 cohomology satisfies $H^*(X;\mathbb{Z}/2\mathbb{Z}) \cong \Lambda(x, Sq^1(x))$, where Λ is the free exterior algebra functor and Sq^1 is the first Steenrod operation. As a CW-complex, $X = P^3(2) \cup e^5$. By Examples 9.7 and 9.9 we obtain: (i) a homotopy equivalence

$$\Omega((S^2 \times S^3)\#X) \simeq \Omega S^2 \times \Omega S^3 \times \Omega W$$

where

$$W = ((\Omega S^2 \times \Omega S^3) \wedge P^3(2)) \vee P^3(2)$$

is homotopy equivalent to a wedge of mod-2 Moore spaces; and (ii) a homotopy fibration

$$U' \vee S^4 \xrightarrow{\Psi} (S^2 \vee S^3) \vee P^3(2) \longrightarrow (S^2 \times S^3)\#X$$

where

$$U' = \Sigma^5\Omega((S^2 \times S^3)\#X)$$

is a wedge of spheres and mod-2 Moore spaces, the restriction of Ψ to U' is a Whitehead product, and the restriction of Ψ to S^4 is the attaching map for the top cell of the connected sum.

CHAPTER 10

Hopf algebras and one-relator algebras

Now that we have many examples of inert maps we take a homological time-out in order to consider the effect an inert map has in homology. To set the stage, consider a homotopy cofibration $\Sigma A \xrightarrow{f} \Sigma Y \xrightarrow{h} Y'$ with the property that Ωh has a right homotopy inverse. Our aim is to calculate the homology of $\Omega Y'$. Take homology with field coefficients. By the Bott-Samelson Theorem there is an algebra isomorphism $H_*(\Omega\Sigma Y) \cong T(\widetilde{H}_*(Y))$, where $T(\ )$ is the free tensor algebra functor.

PROPOSITION 10.1. *Suppose that there is a homotopy cofibration*

$$\Sigma A \xrightarrow{f} \Sigma Y \xrightarrow{h} Y'$$

where Ωh has a right homotopy inverse. Let $\widetilde{f}\colon A \longrightarrow \Omega\Sigma Y$ be the adjoint of f and let $R = \mathrm{Im}(\widetilde{f}_*)$. Then there is an algebra isomorphism

$$H_*(\Omega Y') \cong T(\widetilde{H}_*(Y))/(R)$$

where (R) is the two-sided ideal generated by R.

PROOF. First observe that there is an algebra map $T(\widetilde{H}_*(Y)) \xrightarrow{(\Omega h)_*} H_*(\Omega Y')$. Since Ωh has a right homotopy inverse, $(\Omega h)_*$ is a surjection. Since $\widetilde{f}$ is homotopic to the composite $A \xrightarrow{E} \Omega\Sigma A \xrightarrow{\Omega f} \Omega\Sigma Y$, where E is the suspension, the composite $\Omega h \circ \widetilde{f}$ is null homotopic. Therefore $(\Omega h)_*(R) = 0$. As $(\Omega h)_*$ is an algebra map, we obtain a factorization

$$\begin{array}{ccc} T(\widetilde{H}_*(Y)) & \xrightarrow{(\Omega h)_*} & H_*(\Omega Y') \\ \big\downarrow{\scriptstyle a} & \nearrow{\scriptstyle b} & \\ T(\widetilde{H}_*(Y))/(R) & & \end{array}$$

where a is the quotient map and b is an induced algebra homomorphism. Since $(\Omega h)_*$ is surjective, so is b.

On the other hand, by Theorem 3.5 there is a homotopy fibration

$$\Omega Y' \ltimes \Sigma A \xrightarrow{\chi} \Sigma Y \xrightarrow{h} Y'$$

where χ is the sum of the maps $\Omega Y' \ltimes \Sigma A \xrightarrow{\pi} \Sigma A \xrightarrow{f} \Sigma Y$ and $\Omega Y' \ltimes \Sigma A \xrightarrow{q} \Omega Y' \wedge \Sigma A \xrightarrow{[ev \circ s, f]} \Sigma Y$ for $s\colon \Omega Y' \longrightarrow \Omega Y$ a right homotopy inverse of Ωh. Consider the composite

$$H_*(\Omega(\Omega Y' \ltimes \Sigma A)) \xrightarrow{(\Omega\chi)_*} H_*(\Omega\Sigma Y) \cong T(\widetilde{H}_*(Y)) \xrightarrow{a} T(\widetilde{H}_*(Y))/R.$$

Notice that the maps π and q are suspensions, so the adjoint of $f \circ \pi$ is homotopic to $\alpha\colon \Omega Y' \ltimes A \longrightarrow A \xrightarrow{\widetilde{f}} \Omega\Sigma Y$ and the adjoint of $[ev \circ s, f] \circ q$ is homotopic to $\beta\colon \Omega Y' \ltimes A \longrightarrow \Omega Y' \wedge A \xrightarrow{\langle\widetilde{ev\circ s},\widetilde{f}\rangle} \Omega\Sigma Y$ where the right map is the Samelson product of $\widetilde{ev\circ s}$ (the adjoint of $ev\circ s$) and $\widetilde{f}$. The James construction implies that $\Omega\chi$ is homotopic to the multiplicative extension of $\alpha \perp \beta$. Therefore, as a is an algebra map, $a\circ(\Omega\chi)_*$ is determined by its restriction to $a\circ(\alpha\perp\beta)_*$. By definition, a sends the image of $\widetilde{f}_*$ to the identity element. Therefore $a\circ\alpha_*$ is trivial. Also, the Samelson product commutes with homology in the sense that $(\langle\widetilde{ev\circ s},\widetilde{f}\rangle)_* = \langle(\widetilde{ev\circ s})_*,\widetilde{f}_*\rangle$, where the bracket on the right is the commutator in $T(\widetilde{H}_*(Y))$. The triviality of $a\circ\widetilde{f}_*$ therefore implies that $a\circ(\langle\widetilde{ev\circ s},\widetilde{f}\rangle)_*$ is also trivial. Thus $a\circ\beta_*$ is trivial, implying that $a\circ(\alpha\perp\beta)_*$ is trivial, and hence $a\circ(\Omega\chi)_*$ is trivial.

Further, as the homotopy fibration $\Omega Y'\ltimes\Sigma A\xrightarrow{\chi}\Sigma Y\xrightarrow{h}Y'$ has the property that Ωh has a right homotopy inverse, there is an isomorphism $T(\widetilde{H}_*(Y))\cong H_*(\Omega Y')\otimes H_*(\Omega(\Omega Y'\ltimes\Sigma A))$ of right $H_*(\Omega(\Omega Y'\ltimes\Sigma A))$-modules. Since a is an algebra map and $a\circ(\Omega\chi)_*$ is trivial, we obtain a factorization

$$\begin{array}{ccc} T(\widetilde{H}_*(Y)) & \xrightarrow{(\Omega h)_*} & H_*(\Omega Y') \\ \big\downarrow a & \swarrow c & \\ T(\widetilde{H}_*(Y))/(R) & & \end{array}$$

where c is an algebra map and a surjection.

Finally, consider the composite

$$T(\widetilde{H}_*(Y))/(R)\xrightarrow{b}H_*(\Omega Y')\xrightarrow{c}T(\widetilde{H}_*(Y))/(R)\xrightarrow{b}H_*(\Omega Y').$$

As b and c are surjections, so are $c\circ b$ and $b\circ c$. Therefore $c\circ b$ and $b\circ c$ are surjective self-maps of $T(\widetilde{H}_*(Y))/(R)$ and $H_*(\Omega Y')$ respectively. Any surjective self-map of a graded finite type module is an isomorphism, so both $c\circ b$ and $b\circ c$ are isomorphisms. As b and c are algebra maps, these isomorphisms are as algebras. □

REMARK 10.2. There is an improvement to Proposition 10.1 if Y is a suspension. In that case the Bott-Samelson Theorem improves to a Hopf algebra isomorphism $H_*(\Omega\Sigma Y)\cong T(\widetilde{H}_*(Y))$, where the tensor algebra is primitively generated. The quotient maps b and c in the proof are then Hopf algebra maps, and we obtain an isomorphism of Hopf algebras $H_*(\Omega Y')\cong T(\widetilde{H}_*(Y))/(R)$.

EXAMPLE 10.3. Consider the homotopy cofibration

$$X^{\wedge k}\wedge\Sigma Y\xrightarrow{ad^k(i_1)(i_2)}\Sigma X\vee\Sigma Y\xrightarrow{m_k}M_k.$$

By Lemma 4.12, Ωm_k has a right homotopy inverse. Therefore Proposition 10.1 applies and we obtain an algebra isomorphism

$$H_*(\Omega M_k)\cong T(\widetilde{H}_*(X\vee Y))/(R)$$

where R is the image in homology of the adjoint of $ad^k(i_1)(i_2)$.

A specific case of interest is the homotopy cofibration $S^{km+n+1}\xrightarrow{ad^k(i_1)(i_2)}S^{m+1}\vee S^{n+1}\xrightarrow{m_k}M_k$. We have $T(\widetilde{H}_*(S^m\vee S^n))=T(x,y)$ where $|x|=m$ and $|y|=n$. The adjoint of the iterated Whitehead product $ad^k(i_1)(i_2)$ is an iterated

Samelson product, and its image in homology is the iterated commutator $ad^k(x)(y)$. If $m, n \geq 1$ then by Remark 10.2 there is an isomorphism of Hopf algebras

$$H_*(\Omega M_k) \cong T(x,y)/(ad^k(x)(y)).$$

The special case of Example 10.3 is an example of the notion of a one-relator algebra. In general, an algebra is a *one-relator algebra* if it is not free and can be written as the quotient of a free associative algebra by a two-sided ideal generated by a single element. There are many other examples of one-relator algebras that can be obtained from Proposition 10.1.

EXAMPLE 10.4. Let M be an $(n-1)$-connected $2n$-dimensional Poincaré Duality complex where $n \geq 2$. By Poincaré Duality, as a CW-complex M has one zero-cell, d n-cells for some $d \geq 0$ and one $2n$-cell. If $d = 0$ then $M \simeq S^{2n}$. Otherwise, there is a homotopy cofibration

$$S^{2n-1} \xrightarrow{f} \bigvee_{i=1}^{d} S^n \xrightarrow{h} M$$

where f attaches the top cell to M. In [**BT2**] it was shown that that if $d \geq 2$ then Ωh has a right homotopy inverse. Therefore Proposition 10.1 and Remark 10.2 apply to show that there is an isomorphism of Hopf algebras

$$H_*(\Omega M) \cong T(\widetilde{H}_*(\bigvee_{i=1}^{d} S^{n-1}))/(R)$$

where $R = \mathrm{Im}(\widetilde{f}_*)$. Written explicitly, let $v_i \in H_*(\bigvee_{i=1}^{d} S^{n-1})$ be a generator corresponding to the ith wedge summand of $\bigvee_{i=1}^{d} S^{n-1}$. The image R of $\widetilde{f}_*$ is generated by a single element $r \in T(v_1, \ldots, v_m)$. Therefore there is an isomorphism of Hopf algebras

$$H_*(\Omega M) \cong T(v_1, \ldots, v_m)/(r).$$

A particular example of note is when M is a simply-connected four-manifold.

The following example of a connected sum of products of two simply-connected spheres was calculated in [**GIPS**] using the Adams-Hilton model.

EXAMPLE 10.5. Fix an integer $n \geq 4$. Let $M = \#_{i=1}^{d}(S^{m_i} \times S^{n-m_i})$ where $m_i \geq 2$ for each $1 \leq i \leq d$. Then there is a homotopy cofibration

$$S^{n-1} \xrightarrow{f} \bigvee_{k=1}^{d} S^{m_i} \vee S^{n-m_i} \xrightarrow{h} M$$

where f is the sum of the Whitehead products attaching the top sphere to each copy of $S^{m_i} \times S^{n-m_i}$. Iterating Theorem 9.1 shows that the map Ωh has a right homotopy inverse. Therefore Proposition 10.1 and Remark 10.2 imply that there is a Hopf algebra isomorphism

$$H_*(\Omega M) \cong T(\widetilde{H}_*(\bigvee_{i=1}^{d} S^{m_i-1} \vee S^{n-m_i-1}))/(R)$$

where $R = \mathrm{Im}(\widetilde{f}_*)$. Explicitly, let $u_i \in H_*(S^{m_i-1})$ and $v_i \in H_*(S^{n-m_i-1})$ be generators corresponding to the ith wedge summand in $\bigvee_{i=1}^{d} S^{m_i} \vee S^{n-m_i}$. The image

R of $\widetilde{f}_*$ is then generated by the single element $[u_1, v_1] + \cdots + [u_d, v_d]$. Therefore there is an isomorphism of Hopf algebras

$$H_*(\Omega M) \cong T(u_1, v_1, \ldots, u_d, v_d)/([u_1, v_1] + \cdots + [u_d, v_d]).$$

EXAMPLE 10.6. Let M be an $(n-1)$-connected $(2n+1)$-dimensional Poincaré Duality complex for $n \geq 2$. By Poincaré Duality,

$$H^m(M) \cong \begin{cases} \mathbb{Z} & \text{if } m = 0 \text{ or } m = 2n+1 \\ \mathbb{Z}^d & \text{if } m = n \\ \mathbb{Z}^d \oplus G & \text{if } m = n+1 \\ 0 & \text{otherwise} \end{cases}$$

for some integer $d \geq 0$ and some finite abelian group G. Assume that $d \geq 1$. Let X be the $(n+1)$-skeleton of M. As in [**BT2**], there is a homotopy equivalence $X \simeq (\bigvee_{i=1}^{d}(S^n \vee S^{n+1})) \vee \Sigma V$ where V is a wedge of $(n+1)$-dimensional Moore spaces. Therefore there is a homotopy cofibration

$$S^{2n} \xrightarrow{f} (\bigvee_{i=1}^{d} S^n \vee S^{n+1}) \vee \Sigma V \xrightarrow{i} M$$

By [**BT2**], Ωi has a right homotopy inverse. Thus, by Proposition 10.1 and Remark 10.2 there is an isomorphism of Hopf algebras

$$H_*(\Omega M) \cong T(\widetilde{H}_*((\bigvee_{i=1}^{d} S^{n-1} \vee S^n) \vee V))/(R)$$

where $R = \mathrm{Im}(\widetilde{f})_*$. As in the previous example, this may be rewritten as an isomorphism of Hopf algebras

$$H_*(\Omega M) \cong T(\{u_1, v_1, \ldots, u_d, v_d\} \oplus \widetilde{H}_*(V))/(r)$$

where $|u_i| = n-1$, $|v_i| = n$ and r generates the image of $\widetilde{f}_*$.

REMARK 10.7. Proposition 10.1 does not apply in general to an $(n-1)$-connected $(2n+1)$-dimensional Poincaré Duality complex with $d = 0$. That is, in the case when X is homotopy equivalent to a wedge of Moore spaces. For example, if all the Moore spaces are of the form $P^{n+1}(p^r)$ for a fixed odd prime p and integer r, then there is a homotopy cofibration

$$S^{2n} \xrightarrow{f} \bigvee_{i=1}^{m} P^{n+1}(p^r) \xrightarrow{i} M.$$

We will show that Ωi does not have a right homotopy inverse, implying that one of the hypotheses of Proposition 10.1 fails to hold. By Theorem 7.6 there is a homotopy fibration

$$(\Omega P^{n+1}(p^r) \ltimes \overline{C}) \vee (\bigvee_{i=2}^{m} P^{n+1}(p^r)) \xrightarrow{h} M \xrightarrow{q'} P^{n+1}(p^r)$$

that splits after looping and where $\overline{C} \simeq S^{2n+1} \vee W$ where W is a wedge of mod-p^r Moore spaces. In particular, $\Omega\overline{C}$ is rationally nontrivial (because of the factor ΩS^{2n+1}). However, $\Omega(\bigvee_{i=1}^{m} P^{n+1}(p^r))$ is rationally trivial, so Ωi cannot have a right homotopy inverse. It would be interesting to calculate $H_*(\Omega M)$ in this case.

CHAPTER 11

A second foundational case

This section is in preparation for the next. To set things up, suppose that there is a space M with the property that there is a factorization of the inclusion $\bigvee_{i=1}^{m} \Sigma X_i \longrightarrow \prod_{i=1}^{m} \Sigma X_i$ as a composite

$$\bigvee_{i=1}^{m} \Sigma X_i \xrightarrow{v} M \xrightarrow{w} \prod_{i=1}^{m} \Sigma X_i$$

for some maps v and w. In addition, suppose that there is a homotopy cofibration

$$\Sigma A \xrightarrow{f} M \longrightarrow M'$$

with the property that $w \circ f$ is null homotopic. Then w extends to a map

$$w' \colon M' \longrightarrow \prod_{i=1}^{m} \Sigma X_i$$

and there is a homotopy fibration diagram

$$(11.1) \qquad \begin{array}{ccc} E & \longrightarrow & E' \\ \big\downarrow{\scriptstyle p} & & \big\downarrow{\scriptstyle p'} \\ M & \longrightarrow & M' \\ \big\downarrow{\scriptstyle w} & & \big\downarrow{\scriptstyle w'} \\ \prod_{i=1}^{m} \Sigma X_i & = & \prod_{i=1}^{m} \Sigma X_i \end{array}$$

that defines the spaces E and E' and the maps p and p'. The inclusion $w \circ v$ of the wedge into the product has a right homotopy inverse after looping, implying that Ωw also has a right homotopy inverse $s \colon \prod_{i=1}^{m} \Omega\Sigma X_i \longrightarrow \Omega M$. Theorem 2.2 then implies that there is a homotopy cofibration

$$\prod_{i=1}^{m} \Omega\Sigma X_i \ltimes \Sigma A \xrightarrow{\theta} E \longrightarrow E'$$

and a homotopy commutative diagram

$$\begin{array}{ccc} \prod_{i=1}^{m} \Omega\Sigma X_i \ltimes \Sigma A & \xrightarrow{\theta} & E \\ \big\downarrow{\scriptstyle \simeq} & & \big\downarrow{\scriptstyle p} \\ ((\prod_{i=1}^{m} \Omega\Sigma X_i) \wedge \Sigma A) \vee \Sigma A & \xrightarrow{[\gamma, f] + f} & M \end{array}$$

where γ is the composite $\Sigma(\prod_{i=1}^{m} \Omega\Sigma X_i) \xrightarrow{\Sigma s} \Sigma\Omega M \xrightarrow{ev} M$. On the other hand, the suspension of a product splits as a wedge, and the James construction lets us further split each of the spaces $\Sigma\Omega\Sigma X_i$. In this section we show that those splittings can be chosen so that the maps from the wedge summands into M can be identified as iterated Whitehead products.

Recall from Lemma 3.8 that there is a natural homotopy equivalence

$$\bigvee_{k=1}^{\infty} \Sigma X^{\wedge k} \xrightarrow{\phi} \Sigma\Omega\Sigma X$$

defined as follows. For $k \geq 1$, let e_k be the composite

$$e_k \colon X^{\times k} \xrightarrow{E^{\times k}} (\Omega\Sigma X)^{\times k} \xrightarrow{\mu} \Omega\Sigma X$$

where μ is the standard loop multiplication. There is a natural homotopy equivalence $\Sigma(A \times B) \simeq \Sigma A \vee \Sigma B \vee (\Sigma A \wedge B)$. Iterating this we obtain a natural map $\Sigma X_1 \wedge \cdots \wedge X_k \longrightarrow \Sigma(X_1 \times \cdots \times X_k)$. Let ϕ_k be the composite

$$\phi_k \colon \Sigma X^{\wedge k} \longrightarrow \Sigma(X^{\times k}) \xrightarrow{\Sigma e_k} \Sigma\Omega\Sigma X.$$

Let

$$\phi \colon \bigvee_{k=1}^{\infty} \Sigma X^{\wedge k} \longrightarrow \Sigma\Omega\Sigma X$$

be the wedge sum of the maps ϕ_k for $k \geq 1$. As shorthand, this is called the ϕ-decomposition of $\Sigma\Omega\Sigma X$.

Let $X_1, \ldots, X_m$ be path-connected spaces. For $1 \leq j \leq m$, let

$$t_j \colon X_j \longrightarrow \bigvee_{i=1}^{m} X_i$$

be the inclusion of the jth-wedge summand. Applying the James construction gives a map

$$\Omega\Sigma X_j \xrightarrow{\Omega\Sigma t_j} \Omega\Sigma(\bigvee_{i=1}^{m} X_i).$$

Multiplying the maps $\Omega\Sigma t_j$ together for $1 \leq j \leq m$ gives a map

$$\Psi \colon \prod_{i=1}^{m} \Omega\Sigma X_i \longrightarrow \Omega\Sigma(\bigvee_{i=1}^{m} X_i). \tag{11.2}$$

As Ψ is not $\Omega\Sigma\psi$ for some map ψ, it need not necessarily be the case that the decomposition of $\Sigma(\prod_{i=1}^{m} \Omega\Sigma X_i)$ obtained by combining the natural decomposition of the suspension of a product with the ϕ-decomposition of each $\Sigma\Omega\Sigma X_i$ is compatible with the ϕ-decomposition of $\Sigma\Omega\Sigma(\bigvee_{i=1}^{m} X_i)$. However, in Proposition 11.1 we will show that a decomposition of $\Sigma(\prod_{i=1}^{m} \Omega\Sigma X_i)$ may be chosen to be compatible with the ϕ-decomposition of $\Sigma\Omega\Sigma(\bigvee_{i=1}^{m} X_i)$.

PROPOSITION 11.1. *There is a homotopy equivalence*

$$\bigvee_{k=1}^{\infty} \bigvee_{1 \leq i_1 \leq \cdots \leq i_k \leq m} \Sigma X_{i_1} \wedge \cdots \wedge X_{i_k} \xrightarrow{\varepsilon} \Sigma(\prod_{i=1}^{m} \Omega\Sigma X_i)$$

satisfying a homotopy commutative diagram

$$\begin{array}{ccc} \bigvee_{k=1}^{\infty}\bigvee_{1\le i_1\le\cdots\le i_k\le m}\Sigma X_{i_1}\wedge\cdots\wedge X_{i_k} & \xrightarrow{I} & \bigvee_{k=1}^{\infty}\Sigma(\bigvee_{i=1}^{m} X_i)^{\wedge k} \\ \downarrow{\scriptstyle\varepsilon} & & \downarrow{\scriptstyle\phi} \\ \Sigma(\prod_{i=1}^{m}\Omega\Sigma X_i) & \xrightarrow{\Sigma\Psi} & \Sigma\Omega\Sigma(\bigvee_{i=1}^{m} X_i) \end{array}$$

where I is an inclusion of wedge summands.

PROOF. First consider the diagram

$$\begin{array}{ccccc} X_{i_1}\times\cdots\times X_{i_k} & \xrightarrow{t_{i_1}\times\cdots\times t_{i_k}} & (\bigvee_{i=1}^{m} X_i)^{\times k} & \searrow^{e_k} & \\ \downarrow{\scriptstyle E\times\cdots\times E} & & \downarrow{\scriptstyle E^{\times k}} & & \\ \Omega\Sigma X_{i_1}\times\cdots\times\Omega\Sigma X_{i_k} & \xrightarrow{\Omega\Sigma t_{i_1}\times\cdots\times\Omega\Sigma t_{i_k}} & \Omega\Sigma(\bigvee_{i=1}^{m} X_i)^{\times k} & \xrightarrow{m} & \Omega\Sigma(\bigvee_{i=1}^{m} X_i). \end{array}$$

The left square clearly commutes and the right square commutes by definition of e_k. Now suspend and use the naturality of the map $\Sigma A\wedge B\longrightarrow\Sigma(A\times B)$ to obtain a homotopy commutative diagram

$$\begin{array}{ccccc} \Sigma X_{i_1}\wedge\cdots\wedge X_{i_k} & \xrightarrow{\Sigma t_{i_1}\wedge\cdots\wedge t_{i_k}} & \Sigma(\bigvee_{i=1}^{m} X_i)^{\wedge k} & \searrow^{\phi_k} & \\ \downarrow & & \downarrow & & \\ \Sigma(\Omega\Sigma X_{i_1}\times\cdots\times\Omega\Sigma X_{i_k}) & \xrightarrow{\Sigma(\Omega\Sigma t_{i_1}\times\cdots\times\Omega\Sigma t_{i_k})} & \Sigma\Omega\Sigma(\bigvee_{i=1}^{m} X_i)^{\times k} & \xrightarrow{\Sigma m} & \Sigma\Omega\Sigma(\bigvee_{i=1}^{m} X_i). \end{array}$$

Observe that the map $\Sigma t_{i_1}\wedge\cdots\wedge t_{i_k}$ is the inclusion of a wedge summand. Doing this for each $1\le i_1\le\cdots\le i_k\le m$ then gives a homotopy commutative diagram

$$\begin{array}{ccccc} \bigvee_{1\le i_1\le\cdots\le i_k\le m}\Sigma X_{i_1}\wedge\cdots\wedge X_{i_k} & \xrightarrow{I_k} & \Sigma(\bigvee_{i=1}^{m} X_i)^{\wedge k} & \searrow^{\phi_k} & \\ \downarrow{\scriptstyle\varepsilon_k} & & \downarrow & & \\ \Sigma(\Omega\Sigma X_1\times\cdots\times\Omega\Sigma X_m) & \xrightarrow{\Sigma(\Omega\Sigma t_1\times\cdots\times\Omega\Sigma t_m)} & \Sigma\Omega\Sigma(\bigvee_{i=1}^{m} X_i)^{\times k} & \xrightarrow{\Sigma m} & \Sigma\Omega\Sigma(\bigvee_{i=1}^{m} X_i). \end{array}$$

where I_k is an inclusion of wedge summands. Finally, assembling these diagrams for each $k\ge 1$ gives a homotopy commutative diagram

$$\begin{array}{ccccc} \bigvee_{k=1}^{\infty}\bigvee_{1\le i_1\le\cdots\le i_k\le m}\Sigma X_{i_1}\wedge\cdots\wedge X_{i_k} & \xrightarrow{I} & \bigvee_{k=1}^{\infty}\Sigma(\bigvee_{i=1}^{m} X_i)^{\wedge k} & \searrow^{\phi} & \\ \downarrow{\scriptstyle\varepsilon} & & \downarrow & & \\ \Sigma(\Omega\Sigma X_1\times\cdots\times\Omega\Sigma X_m) & \xrightarrow{\Sigma(\Omega\Sigma t_1\times\cdots\times\Omega\Sigma t_m)} & \Sigma\Omega\Sigma(\bigvee_{i=1}^{m} X_i)^{\times k} & \xrightarrow{\Sigma m} & \Sigma\Omega\Sigma(\bigvee_{i=1}^{m} X_i). \end{array}$$

where I is an inclusion of wedge summands. Observe that the bottom row is $\Sigma\Psi$.

It remains to show that ε is a homotopy equivalence. Take homology with field coefficiets. For $1\le i\le m$, let $V_i=\widetilde{H}_*(X_i)$. By the Bott-Samelson and Kunneth Theorems, there is an algebra isomorphism

$$H_*(\Omega\Sigma X_1\times\cdots\times\Omega\Sigma X_m)\cong T(V_1)\otimes\cdots\otimes T(V_m).$$

The submodule consisting of elements of tensor length k is

$$\bigvee_{1\le i_1\le\cdots\le i_k\le m} V_{i_1}\otimes\cdots\otimes V_{i_k}.$$

Thus the previous isomorphism implies there is a module isomorphism

$$\widetilde{H}_*(\Omega\Sigma X_1\times\cdots\times\Omega\Sigma X_m)\cong\bigvee_{k=1}^{\infty}\bigvee_{1\le i_1\le\cdots\le i_k\le m} V_{i_1}\otimes\cdots\otimes V_{i_k}.$$

For a fixed sequence $(i_1,\ldots,i_k)$ with $1\le i_1\le\cdots\le i_k\le m$, the composite

$$\Sigma X_{i_1}\wedge\cdots\wedge X_{i_k}\longrightarrow\Sigma(X_{i_1}\times\cdots\times X_{i_k})\xrightarrow{\Sigma(E\times\cdots\times E)}\Sigma\Omega\Sigma X_{i_1}\times\cdots\times\Omega\Sigma X_{i_k}$$

induces the inclusion of the submodule $\Sigma V_{i_1}\otimes\cdots\otimes V_{i_k}$. Therefore ε_k induces the submodule inclusion

$$\bigvee_{1\le i_1\le\cdots\le i_k\le m}\Sigma V_{i_1}\otimes\cdots\otimes V_{i_k}\to\widetilde{H}_*(\Sigma(\Omega\Sigma X_1\times\cdots\times\Omega\Sigma X_m),$$

implying that ε induces an isomorphism in homology. As this is true for mod-p homology for all primes p and rational homology, ε incudes an isomorphism in integral homology. Hence it is a homotopy equivalence by Whitehead's Theorem. □

Next is a variation on Proposition 11.1 that involves half-smashes and a generalization of the map c from Chapter 3. The maps b_k in Chapter 3 may be defined more generally as follows. Let

$$\overline{b}_1\colon X_1\wedge\Sigma Y\longrightarrow X_1\ltimes\Sigma Y$$

be the inclusion i. For $k\ge 2$, define

$$\overline{b}_k\colon X_1\wedge\cdots\wedge X_k\wedge\Sigma Y\longrightarrow(X_1\times\cdots\times X_k)\ltimes\Sigma Y$$

recursively by the composite

$$\begin{aligned}X_1\wedge X_2\wedge\cdots\wedge X_k\wedge\Sigma Y&\xrightarrow{\ i\ } X_1\ltimes(X_2\wedge\cdots X_k\wedge\Sigma Y)\\&\xrightarrow{1\ltimes\overline{b}_{k-1}} X_1\ltimes((X_2\times\cdots\times X_k)\ltimes\Sigma Y)\\&\xrightarrow{\ \varphi\ }(X_1\times X_2\times\cdots\times X_k)\ltimes\Sigma Y.\end{aligned}$$

Note that the naturality of i and φ in all variables implies that $\overline{b}_k$ is also natural in all variables. Note also that the map b_k in Chapter 3 is given by taking each X_i for $1\le i\le k$ equal to a common space X. Applying the naturality of $\overline{b}_k$ to the inclusions $X_j\xrightarrow{t_j}\bigvee_{i=1}^m X_i$ then immediately gives the following.

LEMMA 11.2. *Let $X_1,\ldots,X_m$ and Y be path-connected spaces. For any sequence $1\le i_1\le\cdots\le i_k\le m$ there is a homotopy commutative diagram*

$$\begin{array}{ccc}X_{i_1}\wedge\ldots\wedge X_{i_k}\wedge\Sigma Y&\xrightarrow{t_{i_1}\wedge\cdots\wedge t_{i_k}\wedge 1}&(\bigvee_{i=1}^m X_i)^{\wedge k}\wedge\Sigma Y\\ \downarrow{\scriptstyle\overline{b}_k}&&\downarrow{\scriptstyle b_k}\\ (X_{i_1}\times\cdots\times X_{i_k})\ltimes\Sigma Y&\xrightarrow{(t_{i_1}\times\cdots\times t_{i_k})\ltimes 1}&(\bigvee_{i=1}^m X_i)^{\times k}\ltimes\Sigma Y.\end{array}$$

□

In what follows, the $k=0$ case of a smash product $X_{i_1}\wedge\cdots\wedge X_{i_k}\wedge\Sigma A$ refers to ΣA. By Lemma 3.10, there is a homotopy equivalence

$$\bigvee_{k=0}^{\infty}(\vee_{i=1}^{m}X_i)^{\wedge k}\wedge\Sigma A\xrightarrow{c}\Omega\Sigma(\vee_{i=1}^{m}X_i)\ltimes\Sigma A.$$

LEMMA 11.3. *There is a homotopy commutative diagram*

$$\begin{array}{ccc}
\bigvee_{k=0}^{\infty}\ \bigvee_{1\leq i_1\leq\cdots\leq i_k\leq m}(X_{i_1}\wedge\cdots\wedge X_{i_k})\wedge\Sigma A & \xrightarrow{I} & \bigvee_{k=0}^{\infty}(\vee_{i=1}^{m}X_i)^{\wedge k}\wedge\Sigma A\\
\downarrow{\scriptstyle\varepsilon'} & & \downarrow{\scriptstyle c}\\
(\prod_{i=1}^{m}\Omega\Sigma X_i)\ltimes\Sigma A & \xrightarrow{\Psi\ltimes 1} & \Omega\Sigma(\vee_{i=1}^{m}X_i)\ltimes\Sigma A
\end{array}$$

where ε' is a homotopy equivalence and I is an inclusion of wedge summands.

PROOF. The proof is similar to that for Proposition 11.1. It begins with the same first step, just half-smashed with ΣA. Consider the diagram

$$\begin{array}{ccccc}
(X_{i_1}\times\cdots\times X_{i_k})\ltimes\Sigma A & \xrightarrow{(t_{i_1}\times\cdots\times t_{i_k})\ltimes 1} & (\vee_{i=1}^{m}X_i)^{\times k}\ltimes\Sigma A & & \\
\downarrow{\scriptstyle (E\times\cdots\times E)\ltimes 1} & & \downarrow{\scriptstyle E^{\times k}\ltimes 1} & \searrow{\scriptstyle e_k\ltimes 1} & \\
(\Omega\Sigma X_{i_1}\times\cdots\times\Omega\Sigma X_{i_k})\ltimes\Sigma A & \xrightarrow{(\Omega\Sigma t_{i_1}\times\cdots\times\Omega\Sigma t_{i_k})\ltimes 1} & \Omega\Sigma(\vee_{i=1}^{m}X_i)^{\times k}\ltimes\Sigma A & \xrightarrow{m\ltimes 1} & \Omega\Sigma(\vee_{i=1}^{m}X_i)\ltimes\Sigma A.
\end{array}$$

The left square clearly commutes and the right square commutes by definition of e_k.

Next, juxtapose the diagram above with that in Lemma 11.2 (with $Y=A$) to obtain a homotopy commutative diagram

$$\begin{array}{ccccc}
(X_{i_1}\wedge\cdots\wedge X_{i_k})\wedge\Sigma A & \xrightarrow{(t_{i_1}\wedge\cdots\wedge t_{i_k})\wedge 1} & (\vee_{i=1}^{m}X_i)^{\wedge k}\wedge\Sigma A & & \\
\downarrow & & \downarrow & \searrow{\scriptstyle c_k} & \\
(\Omega\Sigma X_{i_1}\times\cdots\times\Omega\Sigma X_{i_k})\ltimes\Sigma A & \xrightarrow{(\Omega\Sigma t_{i_1}\times\cdots\times\Omega\Sigma t_{i_k})\ltimes 1} & \Omega\Sigma(\vee_{i=1}^{m}X_i)^{\times k}\ltimes\Sigma A & \xrightarrow{m\ltimes 1} & \Omega\Sigma(\vee_{i=1}^{m}X_i)\ltimes\Sigma A.
\end{array}$$

Observe that $t_{i_1}\wedge\cdots\wedge t_{i_k}\wedge 1$ is the inclusion of a wedge summand. As in Proposition 11.1, a similar diagram exists for each $1\leq i_1\leq\cdots\leq i_k\leq m$, and then all such diagrams for $k\geq 1$ may be assembled to give a homotopy commutative diagram

$$\begin{array}{ccccc}
\bigvee_{k=1}^{\infty}\ \bigvee_{1\leq i_1\leq\cdots\leq i_k\leq m}\Sigma X_{i_1}\wedge\cdots\wedge X_{i_k}\wedge\Sigma A & \xrightarrow{I} & \vee_{k=1}^{\infty}\Sigma(\vee_{i=1}^{m}X_i)^{\wedge k}\wedge\Sigma A & & \\
\downarrow{\scriptstyle\varepsilon'} & & \downarrow & \searrow{\scriptstyle c} & \\
(\Omega\Sigma X_1\times\cdots\times\Omega\Sigma X_m)\ltimes\Sigma A & \xrightarrow{(\Omega\Sigma t_1\times\cdots\times\Omega\Sigma t_m)\ltimes 1} & \Omega\Sigma(\vee_{i=1}^{m}X_i)^{\times k}\ltimes\Sigma A & \xrightarrow{m\ltimes 1} & \Omega\Sigma(\vee_{i=1}^{m}X_i)\ltimes\Sigma A
\end{array}$$

where I is an inclusion of wedge summands. Observe that the bottom row is $\Psi\ltimes 1$ so the homotopy commutativity of the diagram implies that $c\circ I\simeq(\Psi\ltimes 1)\circ\varepsilon'$. An argument as in Proposition 11.1 shows that ε' is a homotopy equivalence. □

Recall from the setup at the beginning of the section that there is a composite $\vee_{i=1}^{m}\Sigma X_i\xrightarrow{v}M\xrightarrow{w}\prod_{i=1}^{m}\Sigma X_i$ that is homotopic to the inclusion of the wedge into the product. For $1 \leq k \leq m$, let v_k be the composite

$$v_k \colon \Sigma X_k \xrightarrow{\Sigma t_k} \bigvee_{i=1}^{m} \Sigma X_i \xrightarrow{v} M.$$

Recall as well that two maps f and g are congruent if $\Sigma f \simeq \Sigma g$, implying that $f_* = g_*$.

THEOREM 11.4. *There is a homotopy cofibration*

$$\bigvee_{k=0}^{\infty} \bigvee_{1\leq i_1\leq\cdots\leq i_k\leq m} (X_{i_1} \wedge \cdots \wedge X_{i_k}) \wedge \Sigma A \xrightarrow{\zeta} E \longrightarrow E'$$

where the map ζ is congruent to a map ζ' satisfying a homotopy commutative diagram

$$\begin{array}{ccc}
\bigvee_{k=0}^{\infty} \bigvee_{1\leq i_1\leq\cdots\leq i_k\leq m} (X_{i_1} \wedge \cdots \wedge X_{i_k}) \wedge \Sigma A & \xrightarrow{\zeta'} & E \\
 & \searrow^{\bigvee_{k=0}^{\infty} \bigvee_{1\leq i_1\leq\cdots\leq i_k\leq m} [v_{i_1},[v_{i_2},[\cdots[v_{i_k},f]]\cdots]} & \downarrow{\scriptstyle p} \\
 & & M.
\end{array}$$

PROOF. The proof is broken into steps.

STEP 1 (setting up). After looping, the inclusion $\bigvee_{i=1}^{m} \Sigma X_i \longrightarrow \prod_{i=1}^{m} \Sigma X_i$ has a right homotopy inverse. A specific choice of a right homotopy inverse is given by the map Ψ defined in (11.2). Let s be the composite

$$s \colon \prod_{i=1}^{m} \Omega\Sigma X_i \xrightarrow{\Psi} \Omega\Sigma(\bigvee_{i=1}^{m} X_i) \xrightarrow{\Omega v} \Omega M.$$

Then as $w \circ v$ is homotopic to the inclusion of the wedge into the product, s is a right homotopy inverse for Ωw.

As an intermediate stage, define the space $\overline{E}$ and the map $\overline{p}$ by the homotopy fibration

$$\overline{E} \xrightarrow{\overline{p}} (\bigvee_{i=1}^{m} \Sigma X_i) \vee \Sigma A \xrightarrow{q_1} \bigvee_{i=1}^{m} \Sigma X_i$$

where q_1 is the pinch map. By Example 3.6 there is a lift

$$\overline{g} \colon \Sigma A \longrightarrow \overline{E}$$

of the inclusion $\Sigma A \xrightarrow{i_2} (\bigvee_{i=1}^{m} \Sigma X_i) \vee \Sigma A$ and a homotopy equivalence

$$\Omega\Sigma(\bigvee_{i=1}^{m} \Sigma X_i) \ltimes \Sigma A \xrightarrow{\overline{a}\circ(1\ltimes\overline{g})} \overline{E}.$$

Since w extends the inclusion of the wedge into the product and $w \circ f$ is null homotopic, there is a homotopy commutative square

$$\begin{array}{ccc}
(\bigvee_{i=1}^{m} \Sigma X_i) \vee \Sigma A & \xrightarrow{v\perp f} & M \\
\downarrow{\scriptstyle q_1} & & \downarrow{\scriptstyle w} \\
\bigvee_{i=1}^{m} \Sigma X_i & \longrightarrow & \prod_{i=1}^{m} \Sigma X_i.
\end{array}$$

Let $\alpha\colon \overline{E} \longrightarrow E$ be the induced map of fibres and let g be the composite

$$g\colon \Sigma A \xrightarrow{\overline{g}} \overline{E} \xrightarrow{\alpha} E.$$

Notice that as $\overline{g}$ is a lift of the identity map on ΣA to $\overline{E}$, the map g is a lift of f to E. Further, we claim that the composite $\Sigma A \xrightarrow{g} E \longrightarrow E'$ is null homotopic. Since g lifts f, the composite $\Sigma A \xrightarrow{g} E \longrightarrow E' \xrightarrow{p'} M'$ is homotopic to $\Sigma A \xrightarrow{f} M \longrightarrow M'$ by (11.1), which is null homotopic since M' is the homotopy cofibre of f. Thus $\Sigma A \xrightarrow{g} E \longrightarrow E'$ lifts to the homotopy fibre of p'. But by (11.1), as Ωw has a right homotopy inverse, so does $\Omega w'$, implying that the connecting map for the homotopy fibration $E' \xrightarrow{p'} M' \xrightarrow{w'} \prod_{i=1}^{m} \Sigma X_i$ is null homotopic. Hence the lift of $\Sigma A \xrightarrow{g} E \longrightarrow E'$ to the homotopy fibre of p' must be null homotopic, implying that the composite $\Sigma A \xrightarrow{g} E \longrightarrow E'$ is null homotopic.

With this choice of g and the existence of a right homotopy inverse for Ωw, Theorem 2.2 implies that there is a homotopy cofibration

$$(\prod_{i=1}^{m} \Omega\Sigma X_i) \ltimes \Sigma A \xrightarrow{\theta} E \longrightarrow E' \tag{11.3}$$

where, by definition, θ is the composite

$$(\prod_{i=1}^{m} \Omega\Sigma X_i) \ltimes \Sigma A \xrightarrow{s \ltimes g} \Omega M \ltimes E \xrightarrow{\Gamma} E.$$

STEP 2 (the map ζ). Consider the diagram
(11.4)

$$\begin{array}{ccccc}
\bigvee_{k=0}^{\infty} \bigvee_{1\leq i_1\leq\cdots\leq i_k\leq m} (X_{i_1}\wedge\cdots\wedge X_{i_k})\wedge\Sigma A & \xrightarrow{\varepsilon'} & (\prod_{i=1}^{m}\Omega\Sigma X_i)\ltimes\Sigma A & & \\
\downarrow I & & \downarrow \Psi\ltimes 1 & \searrow s\ltimes g & \\
\bigvee_{k=0}^{\infty}(\bigvee_{i=1}^{m} X_i)^{\wedge k}\wedge\Sigma A & \xrightarrow{c} & \Omega\Sigma(\bigvee_{i=1}^{m} X_i)\ltimes\Sigma A & \xrightarrow{\Omega v\ltimes g} & \Omega M\ltimes E \xrightarrow{\Gamma} E.
\end{array}$$

The left square homotopy commutes by Lemma 11.3 and the triangle homotopy commutes by definition of s. In the upper direction around the diagram, $\Gamma\circ(s\ltimes g)$ is the definition of θ and, by Lemma 11.3, ε' is a homotopy equivalence. So if $\zeta = \theta\circ\varepsilon'$ then by (11.3) there is a homotopy cofibration

$$\bigvee_{k=0}^{\infty} \bigvee_{1\leq i_1\leq\cdots\leq i_k\leq m} (X_{i_1}\wedge\cdots\wedge X_{i_k})\wedge\Sigma A \xrightarrow{\zeta} E \longrightarrow E'.$$

STEP 3 (the map ζ' and the congruence with ζ). In the lower row of (11.4) inserting the homotopy equivalence

$$\beta\colon \Omega\Sigma(\bigvee_{i=1}^{m} \Sigma X_i) \ltimes \Sigma A \xrightarrow{\overline{a}\circ(1\ltimes\overline{g})} \overline{E}$$

and its inverse gives $\Gamma\circ(\Omega v\ltimes g)\circ c \simeq \Gamma\circ(\Omega v\ltimes g)\circ\beta^{-1}\circ\overline{a}\circ(1\ltimes\overline{g})\circ c$. Notice that the composite $\overline{a}\circ(1\ltimes\overline{g})\circ c$ is the map d from Chapter 3. By Theorem 3.15,

d is congruent to a map $\mathfrak{d}$ that satisfies a homotopy commutative diagram

$$\begin{array}{ccc} \bigvee_{k=0}^{\infty}(\bigvee_{i=1}^{m} X_i)^{\wedge k} \wedge \Sigma A & \xrightarrow{\quad\mathfrak{d}\quad} & \overline{E} \\ & \searrow^{\bigvee_{k=1}^{\infty} ad^k(i_W)(i_{\Sigma A})} & \big\downarrow \overline{p} \\ & & (\bigvee_{i=1}^{m} \Sigma X_i) \vee \Sigma A \end{array} \tag{11.5}$$

where i_W and $i_{\Sigma A}$ are the inclusions of $\bigvee_{i=1}^{m} \Sigma X_i$ and ΣA respectively into

$$(\bigvee_{i=1}^{m} \Sigma X_i) \vee \Sigma A.$$

Let

$$\zeta' = \Gamma \circ (\Omega v \ltimes g) \circ \beta^{-1} \circ \mathfrak{d} \circ I.$$

The congruence between $\mathfrak{d}$ and $d = \overline{a} \circ (1 \ltimes \overline{g}) \circ c$ implies that ζ' is congruent to $\Gamma \circ (\Omega v \ltimes g) \circ \beta^{-1} \circ d \circ I$. As $\beta = \overline{a} \circ (1 \ltimes \overline{g})$, we obtain a congruence between ζ' and $\Gamma \circ (\Omega v \ltimes g) \circ c \circ I$. The latter is the lower direction around the diagram (11.4), and so is homotopic to the upper direction around that diagram, which is $\theta \circ \varepsilon' = \zeta$. Hence ζ' is congruent to ζ.

STEP 4 (identifying Whitehead products). Finally, consider the diagram

$$\begin{array}{ccccccccc} \bigvee_{k=0}^{\infty}(\bigvee_{i=1}^{m} X_i)^{\wedge k} \wedge \Sigma A & \xrightarrow{\mathfrak{d}} & \overline{E} & \xrightarrow{\beta^{-1}} & \Omega\Sigma(\bigvee_{i=1}^{m} X_i) \ltimes \Sigma A & \xrightarrow{\Omega v \ltimes g} & \Omega M \ltimes E & \xrightarrow{\Gamma} & E \\ & \searrow^{\bigvee_{k=1}^{\infty} ad^k(i_W)(i_{\Sigma A})} & \big\downarrow \overline{p} & & \big\downarrow & & \big\downarrow & & \big\downarrow p \\ & & (\bigvee_{i=1}^{m} \Sigma X_i) \vee \Sigma A & = & (\bigvee_{i=1}^{m} \Sigma X_i) \vee \Sigma A & \xrightarrow{v \vee g} & M \vee E & \xrightarrow{1 \vee p} & M. \end{array}$$

The left triangle homotopy commutes by (11.5), the left of centre square homotopy commutes by the homotopy equivalence β, the right of centre square homotopy commutes by naturality and the right square commutes by the definition of Γ in Chapter 2. The upper direction around the diagram, precomposed with I, is the definition of ζ'. The lower direction around the diagram, precomposed with I, behaves as follows. Observe that the restriction of I to the wedge summand $X_{i_1} \wedge \cdots \wedge X_{i_k} \wedge \Sigma A$ is the inclusion $t_{i_1} \wedge \cdots \wedge t_{i_k} \wedge 1$. Thus the restriction of $ad^k(i_W)(i_{\Sigma A})$ to $X_{i_1} \wedge \cdots \wedge X_{i_k} \wedge \Sigma A$ is $[\Sigma t_{i_1}, [\Sigma t_{i_2}, [\cdots [\Sigma t_{i_k}, i_{\Sigma A}]] \cdots]$. The naturality of the Whitehead product, the definition of v_k as $v \circ \Sigma\iota_k$, and the fact that $p \circ g \simeq f$ imply that

$$(v \vee (p \circ g)) \circ [\Sigma\iota_{i_1}, [\Sigma\iota_{i_2}, [\cdots [\Sigma\iota_{i_k}, i_{\Sigma A}]] \cdots] \simeq [v_{i_1}, [v_{i_2}, [\cdots [v_{i_k}, f]] \cdots].$$

Thus the lower direction around the diagram is the wedge sum of the iterated Whitehead products $[v_{i_1}, [v_{i_2}, [\cdots [v_{i_k}, f]] \cdots]$ for all $1 \leq i_1 \leq \cdots \leq i_k \leq m$ and all $k \geq 0$. Hence

$$p \circ \zeta' \simeq \bigvee_{k=0}^{\infty} \bigvee_{1 \leq i_1 \leq \cdots \leq i_k \leq m} [v_{i_1}, [v_{i_2}, [\cdots [v_{i_k}, f]] \cdots]$$

as asserted.

∎

CHAPTER 12

Polyhedral products and Whitehead products

The main application of Theorem 11.4 is to polyhedral products. We first recall and formalize the definition in the Introduction. Let K be an abstract simplicial complex on the vertex set $[m] = \{1, 2, \ldots, m\}$. That is, K is a collection of subsets $\sigma \subseteq [m]$ such that for any $\sigma \in K$ all subsets of σ also belong to K. We will usually refer to K as a simplicial complex rather than an abstract simplicial complex. A subset $\sigma \in K$ is a *simplex* or *face* of K. The emptyset $\emptyset$ is assumed to belong to K. For $1 \leq i \leq m$, let (X_i, A_i) be a pair of pointed CW-complexes, where A_i is a pointed subspace of X_i. Let $(\underline{X}, \underline{A}) = \{(X_i, A_i)\}_{i=1}^m$ be the sequence of CW-pairs. For each face $\sigma \in K$, let $(\underline{X}, \underline{A})^\sigma$ be the subspace of $\prod_{i=1}^m X_i$ defined by

$$(\underline{X}, \underline{A})^\sigma = \prod_{i=1}^{m} Y_i \qquad \text{where} \qquad Y_i = \begin{cases} X_i & \text{if } i \in \sigma \\ A_i & \text{if } i \notin \sigma. \end{cases}$$

The *polyhedral product* determined by $(\underline{X}, \underline{A})$ and K is

$$(\underline{X}, \underline{A})^K = \bigcup_{\sigma \in K} (\underline{X}, \underline{A})^\sigma \subseteq \prod_{i=1}^{m} X_i.$$

For example, suppose each A_i is a point. If K is a disjoint union of m points then $(\underline{X}, \underline{*})^K$ is the wedge $X_1 \vee \cdots \vee X_m$, and if K is the standard $(m-1)$-simplex then $(\underline{X}, \underline{*})^K$ is the product $X_1 \times \cdots \times X_m$.

We aim to apply Theorem 11.4 in the context of a homotopy cofibration $\Sigma A \longrightarrow (\underline{\Sigma X}, \underline{*})^K \longrightarrow (\underline{\Sigma X}, \underline{*})^{\overline{K}}$; this will be done in Proposition 12.6. To prepare some definitions and preliminary results are needed.

The *boundary* of a simplex σ, written $\partial\sigma$, is the simplicial complex consisting of all the proper subsets of σ. A simplex σ is a (minimal) *missing face* of K if $\sigma \notin K$ but $\partial\sigma \subseteq K$. The geometric realization of K is written $|K|$. Note that if σ is a face of K with k elements then $|\sigma| \cong \Delta^{k-1}$, and $|\partial\sigma| \cong \partial\Delta^{k-1}$. The *dimension* of K, written $\dim(|K|)$, is the dimension of the geometric realization $|K|$.

Given a simplicial complex K on the vertex set $[m]$, let $\mathcal{S} = \{\sigma_1, \ldots, \sigma_r\}$ be a subset of the set of missing faces of K. Define a new simplicial complex $\overline{K}$ by

$$\overline{K} = K \cup \mathcal{S}.$$

In terms of geometric realizations, $|\overline{K}|$ is obtained from $|K|$ by taking the missing faces indexed by $\mathcal{S}$ and gluing them to $|K|$ along their boundaries. The naturality of the polyhedral product implies that the simplicial inclusion $K \longrightarrow \overline{K}$ induces a map $(\underline{X}, \underline{A})^K \longrightarrow (\underline{X}, \underline{A})^{\overline{K}}$.

We now specialize to pairs of the form $(X_i, *)$ in order to better identify certain spaces. By definition of the polyhedral product we have

$$(\underline{X}, \underline{*})^{\Delta^{m-1}} = \prod_{i=1}^{m} X_i.$$

The *fat wedge* is the subspace of $\prod_{i=1}^{m} X_i$ defined by

$$FW(X_1, \ldots, X_m) = \{(x_1, \ldots, x_m) \in \prod_{i=1}^{m} X_i \mid \text{at least one } x_i \text{ is } *\}.$$

The definition of the polyhedral product implies that

$$(\underline{X}, \underline{*})^{\partial\Delta^{m-1}} = FW(X_1, \ldots, X_m).$$

Thus if $\sigma = (i_1, \ldots, i_k) \subseteq [m]$ then

$$(\underline{X}, \underline{*})^{\sigma} = \prod_{j=1}^{k} X_{i_j} \qquad \text{and} \qquad (\underline{X}, \underline{*})^{\partial\sigma} = FW(X_{i_1}, \ldots, X_{i_k}).$$

Therefore, in our case, for each missing face $\sigma = (i_1, \ldots, i_k) \in \mathcal{S}$ there is a cofibration

$$FW(X_{i_1}, \ldots, X_{i_k}) \longrightarrow \prod_{j=1}^{k} X_{i_j} \longrightarrow X_{i_1} \wedge \cdots \wedge X_{i_k}. \tag{12.1}$$

We show that an analogue is true for the map of polyhedral products $(\underline{X}, \underline{*})^K \longrightarrow (\underline{X}, \underline{*})^{\overline{K}}$.

REMARK 12.1. It is worth pointing out in what follows that when we write $\bigvee_{\sigma \in \mathcal{S}} X_{i_1} \wedge \cdots \wedge X_{i_k}$ we mean $\sigma = (i_1, \ldots, i_k)$ and it is understood that the number of vertices k may be different for distinct missing faces in $\mathcal{S}$.

LEMMA 12.2. *Suppose that for $1 \leq i \leq m$ each space X_i is path-connected and each missing face in $\mathcal{S}$ has at least two vertices. Then there is a homotopy cofibration*

$$(\underline{X}, \underline{*})^K \longrightarrow (\underline{X}, \underline{*})^{\overline{K}} \longrightarrow \bigvee_{\sigma \in \mathcal{S}} X_{i_1} \wedge \cdots \wedge X_{i_k}.$$

PROOF. Define the space C by the cofibration

$$(\underline{X}, \underline{*})^K \longrightarrow (\underline{X}, \underline{*})^{\overline{K}} \longrightarrow C.$$

In general, if L is a simplicial complex on the vertex set $[m]$ then by [**BBCG**] $\Sigma(\underline{X}, \underline{*})^L$ is homotopy equivalent to $\bigvee_{\tau \in L} \Sigma X_{i_1} \wedge \cdots \wedge X_{i_\ell}$ where $\tau = (i_1, \ldots, i_\ell)$, and this decomposition is natural with respect to simplicial maps $L \longrightarrow L'$. In our case, as $\overline{K}$ consists of all the faces of K together with the missing faces indexed by $\mathcal{S}$, we obtain $\Sigma C \simeq \bigvee_{\sigma \in \mathcal{S}} \Sigma X_{i_1} \wedge \cdots \wedge X_{i_k}$. We claim that this decomposition for ΣC desuspends.

Fix $\sigma = (i_1, \ldots, i_k) \in \mathcal{S}$. Consider the diagram

$$\begin{array}{ccccc}
FW(X_{i_1}, \ldots, X_{i_k}) & \longrightarrow & \prod_{j=1}^{k} X_{i_j} & \longrightarrow & X_{i_1} \wedge \cdots \wedge X_{i_k} \\
\downarrow & & \downarrow & & \downarrow{\scriptstyle g_\sigma} \\
(\underline{X}, \underline{*})^K & \longrightarrow & (\underline{X}, \underline{*})^{\overline{K}} & \longrightarrow & C
\end{array}$$

where the map g_σ will be defined momentarily. Since σ is a missing face for K but is a face of $\overline{K}$, the full subcomplexes of $(\underline{X}, \underline{A})^K$ and $(\underline{X}, \underline{A})^{\overline{K}}$ on the vertex set $\{i_1, \ldots, i_k\}$ are $FW(X_{i_1}, \ldots, X_{i_k})$ and $\prod_{j=1}^k X_{i_j}$ respectively. Therefore, by the naturality of the polyhedral product with respect to simplicial maps, the left square above commutes. This induces a map of cofibres, which gives the right square and defines g_σ. Notice that the decomposition of ΣC implies that Σg_σ is the inclusion of a wedge summand. Thus if

$$g\colon \bigvee_{\sigma \in \mathcal{S}} X_{i_1} \wedge \cdots \wedge X_{i_k} \longrightarrow C$$

is the wedge sum of the maps g_σ for all $\sigma \in \mathcal{S}$, then Σg is a homotopy equivalence. This implies that g_* induces an isomorphism in homology. As each space X_i is path-connected and we assume that each missing face in $\mathcal{S}$ has at least two vertices, the spaces $X_{i_1} \wedge \cdots \wedge X_{i_k}$ are simply-connected. Hence, by Whitehead's Theorem, g_* inducing an isomorphism in homology implies that g is a homotopy equivalence. □

REMARK 12.3. A useful piece of information to record from the proof of Lemma 12.2 is that if $\sigma \in \mathcal{S}$ then the inclusion of σ into $\overline{K}$ induces a homotopy cofibration diagram

$$\begin{array}{ccccc} FW(X_{i_1}, \ldots, X_{i_k}) & \longrightarrow & \prod_{j=1}^k X_{i_j} & \longrightarrow & X_{i_1} \wedge \cdots \wedge X_{i_k} \\ \downarrow & & \downarrow & & \downarrow{\scriptstyle g_\sigma} \\ (\underline{X}, \underline{*})^K & \longrightarrow & (\underline{X}, \underline{*})^{\overline{K}} & \longrightarrow & \bigvee_{\sigma \in \mathcal{S}} X_{i_1} \wedge \cdots \wedge X_{i_k} \end{array}$$

where g_σ is the inclusion of a wedge summand.

We now specialize further to pairs of the form $(\Sigma X_i, *)$ in order to turn the cofibration in Lemma 12.2 one step to the left. The (reduced) *join* of two pointed spaces A and B is the quotient space $A * B = (A \times I \times B)/\sim$ where $I = [0,1]$ is the unit interval and the defining relations are given by $(a, 1, b) \sim (a', 1, b)$, $(a, 0, b) \sim (a, 0, b')$ and $(*, t, *) \sim (*, 0, *)$ for all $a, a' \in A$, $b, b' \in B$ and $t \in I$. There is a well known homotopy equivalence $A * B \simeq \Sigma A \wedge B$.

In the case of pairs $(\Sigma X_i, *)$, for each missing face $\sigma = (i_1, \ldots, i_k) \in \mathcal{S}$ there is a homotopy cofibration

$$X_{i_1} * \cdots * X_{i_k} \longrightarrow FW(\Sigma X_{i_1}, \ldots, \Sigma X_{i_k}) \longrightarrow \prod_{j=1}^k \Sigma X_{i_j}$$

that induces the cofibration in (12.1). Let

$$f\colon \bigvee_{\sigma \in \mathcal{S}} X_{i_1} * \cdots * X_{i_k} \longrightarrow (\Sigma \underline{X}, \underline{*})^K$$

be the wedge sum of the composites $X_{i_1} * \cdots * X_{i_k} \longrightarrow FW(\Sigma X_{i_1}, \ldots, \Sigma X_{i_k}) \longrightarrow (\Sigma \underline{X}, \underline{*})^K$ for all $\sigma = (i_1, \ldots, i_k) \in \mathcal{S}$.

LEMMA 12.4. *Suppose that each missing face in $\mathcal{S}$ has at least two vertices. Then there is a homotopy cofibration*

$$\bigvee_{\sigma \in \mathcal{S}} X_{i_1} * \cdots * X_{i_k} \xrightarrow{f} (\Sigma \underline{X}, \underline{*})^K \longrightarrow (\Sigma \underline{X}, \underline{*})^{\overline{K}}$$

that induces the homotopy cofibration in Lemma 12.2.

PROOF. In general, as $\overline{K} = K \cup S$, the definition of the polyhedral product implies that there is a pushout

$$\begin{array}{ccc} \bigcup_{\sigma\in\mathcal{S}} FW(X_{i_1},\ldots,X_{i_k}) & \longrightarrow & \bigcup_{\sigma\in\mathcal{S}}\left(\prod_{j=1}^{k} X_{i_j}\right) \\ \downarrow & & \downarrow \\ (\underline{X},\underline{*})^K & \longrightarrow & (\underline{X},\underline{*})^{\overline{K}}. \end{array} \tag{12.2}$$

By Lemma 12.2, the homotopy cofibre along the bottom row is $\bigvee_{\sigma\in\mathcal{S}} X_{i_1}\wedge\cdots\wedge X_{i_k}$. The fact that (12.2) is a pushout implies that the cofibre of the top row is also $\bigvee_{\sigma\in\mathcal{S}} X_{i_1}\wedge\cdots\wedge X_{i_k}$. By Remark 12.3, the restriction of (12.2) to

$$FW(X_{i_1},\ldots,X_{i_k}) \longrightarrow \prod_{j=1}^{k} X_{i_j}$$

for a fixed σ induces the inclusion of the wedge summand $X_{i_1}\wedge\cdots\wedge X_{i_k}$ into the cofibre. In our case each such map $FW(\Sigma X_{i_1},\ldots,\Sigma X_{i_k}) \longrightarrow \prod_{j=1}^{k}\Sigma X_{i_j}$ is induced by a map $X_{i_1}*\cdots*X_{i_k} \longrightarrow FW(\Sigma X_{i_1},\ldots,\Sigma X_{i_k})$. Therefore there is a homotopy cofibration sequence

$$\begin{aligned} \bigvee_{\sigma\in S} X_{i_1}*\cdots*X_{i_k} &\longrightarrow \bigcup_{\sigma\in S} FW(\Sigma X_{i_1},\ldots,\Sigma X_{i_k}) \\ &\longrightarrow \bigcup_{\sigma\in S}\left(\prod_{j=1}^{k}\Sigma X_{i_j}\right) \\ &\longrightarrow \bigvee_{\sigma\in S}\Sigma X_{i_1}\wedge\cdots\wedge\Sigma X_{i_k}. \end{aligned}$$

Hence, as (12.2) is a pushout, the definition of f implies that there is a homotopy cofibration

$$\bigvee_{\sigma\in\mathcal{S}} X_{i_1}*\cdots*X_{i_k} \xrightarrow{f} (\Sigma\underline{X},\underline{*})^K \longrightarrow (\Sigma\underline{X},\underline{*})^{\overline{K}}. \qquad \square$$

REMARK 12.5. Lemma 12.4 is also proved in [**IK2**, Theorem 5.1 and Remark 5.2] as a consequence of a grand organizational scheme for polyhedral products called the fat wedge filtration. Our formulation is more elementary as the focus is only on what is needed for Lemma 12.4.

Observe that $X_{i_1}*\cdots*X_{i_k}\simeq\Sigma^{k-1}X_{i_1}\wedge\cdots\wedge X_{i_k}$. As we assume each missing face $\sigma=(i_1,\ldots,i_k)\in\mathcal{S}$ has at least two vertices, we have $k\geq 2$, implying that $\Sigma^{k-1}X_{i_1}\wedge\cdots\wedge X_{i_k}$ is a suspension. Let

$$A = \bigvee_{\sigma\in\mathcal{S}}\Sigma^{k-2}X_{i_1}\wedge\cdots\wedge X_{i_k}.$$

As in Remark 12.1, note that the number k depends on σ and may be different for distinct elements of $\mathcal{S}$. The homotopy cofibration in Lemma 12.4 may now be rewritten as follows.

PROPOSITION 12.6. *Let K be a simplicial complex on the vertex set $[m]$, let $\mathcal{S}$ be a subset of the missing faces of K, and let $\overline{K} = K\cup\mathcal{S}$. Then there is a homotopy cofibration*

$$\Sigma A \longrightarrow (\Sigma\underline{X},\underline{*})^K \longrightarrow (\Sigma\underline{X},\underline{*})^{\overline{K}}.$$

□

The point of Proposition 12.6 is to put us in a position to apply Theorem 11.4. Let K be a simplicial complex on the vertex set $[m]$, let $\mathcal{S}$ be a subset of the missing faces of K and let $\overline{K} = K \cup \mathcal{S}$. Then there is a homotopy fibration diagram

$$\begin{array}{ccc} E & \longrightarrow & \overline{E} \\ \big\downarrow{\scriptstyle p} & & \big\downarrow{\scriptstyle \overline{p}} \\ (\underline{\Sigma X}, \underline{*})^K & \longrightarrow & (\underline{\Sigma X}, \underline{*})^{\overline{K}} \\ \big\downarrow{\scriptstyle w} & & \big\downarrow{\scriptstyle \overline{w}} \\ \prod_{i=1}^m \Sigma X_i & = & \prod_{i=1}^m \Sigma X_i \end{array}$$

where w and $\overline{w}$ are inclusions. By [**DS**], there is are homotopy equivalences

$$E \simeq (\underline{C\Omega\Sigma X}, \underline{\Omega\Sigma X})^K \qquad \overline{E} \simeq (\underline{C\Omega\Sigma X}, \underline{\Omega\Sigma X})^{\overline{K}}$$

and, under these equivalences, the maps p and $\overline{p}$ become maps of polyhedral products induced by appropriate maps of pairs of spaces. The inclusion of the vertex set into K induces a map of polyhedral products

$$v\colon \bigvee_{i=1}^m \Sigma X_i \longrightarrow (\underline{\Sigma X}, \underline{*})^K$$

with the property that the composite $\bigvee_{i=1}^m \Sigma X_i \xrightarrow{v} (\underline{\Sigma X}, \underline{*})^K \xrightarrow{w} \prod_{i=1}^m \Sigma X_i$ is the inclusion of the wedge into the product. For $1 \leq k \leq m$, let v_k be the composite

$$v_k\colon \Sigma X_k \xrightarrow{\Sigma t_k} \bigvee_{i=1}^m \Sigma X_i \xrightarrow{v} (\underline{\Sigma X}, \underline{*})^K.$$

By Lemma 12.6 there is a homotopy cofibration

$$\Sigma A \longrightarrow (\underline{\Sigma X}, \underline{*})^K \longrightarrow (\underline{\Sigma X}, \underline{*})^{\overline{K}}.$$

Thus all the hypotheses of Theorem 11.4 apply and we obtain the following, which is a restatement of Theorem 1.5.

THEOREM 12.7. *There is a homotopy cofibration*

$$\bigvee_{k=0}^{\infty} \bigvee_{1 \leq i_1 \leq \cdots \leq i_k \leq m} (X_{i_1} \wedge \cdots \wedge X_{i_k}) \wedge \Sigma A \xrightarrow{\zeta} (\underline{C\Omega\Sigma X}, \underline{\Omega\Sigma X})^K \longrightarrow (\underline{C\Omega\Sigma X}, \underline{\Omega\Sigma X})^{\overline{K}}$$

where the map ζ is congruent to a map ζ' satisfying a homotopy commutative diagram

$$\begin{array}{ccc}
\bigvee_{k=0}^{\infty} \bigvee_{1\leq i_1\leq\cdots\leq i_k\leq m} (X_{i_1}\wedge\cdots\wedge X_{i_k})\wedge\Sigma A & \xrightarrow{\zeta'} & (\underline{C\Omega\Sigma X}, \underline{\Omega\Sigma X})^K \\
& \searrow^{\bigvee_{k=1}^{\infty}\bigvee_{1\leq i_1\leq\cdots\leq i_k\leq m}[v_{i_1},[v_{i_2},[\cdots[v_{i_k},f]]\cdots]} & \big\downarrow p \\
& & (\underline{\Sigma X}, \underline{*})^K.
\end{array}$$

□

The value of Theorem 12.7 comes from the potential for playing off ζ and ζ' in order to determine the homotopy type of $(\underline{C\Omega\Sigma X}, \underline{\Omega\Sigma X})^{\overline{K}}$ or the homotopy class of p. One way this can be used is explored in the next subsection. Before beginning that, it is worth noting that there are many contexts in which one might examine the transition in polyhedral products from $(\underline{C\Omega\Sigma X}, \underline{\Omega\Sigma X})^K$ to $(\underline{C\Omega\Sigma X}, \underline{\Omega\Sigma X})^{\overline{K}}$.

EXAMPLE 12.8. One way to form a simplicial complex is to start with the vertex set and iteratively add one missing face at a time. For example, if $\sigma = (i_1, \ldots, i_\ell)$ is a missing face of K and $\overline{K} = K \cup \sigma$ then the space ΣA in Theorem 12.7 is $X_{i_1} * \cdots * X_{i_\ell}$, and the theorem informs on the homotopy type of $(\underline{C\Omega\Sigma X}, \underline{\Omega\Sigma X})^{\overline{K}}$.

EXAMPLE 12.9. The process in Example 12.8 may be accelerated by building up the simplicial complex skeleton-by-skeleton. Let K be a simplicial complex. For $0 \leq t \leq m-1$ let K_t be the full t-skeleton of K. That is, K_t is the simplicial complex consisting of all the faces of K of dimension $\leq t$. Notice that if $\sigma \in K_t$ then $\partial\sigma \subseteq K_{t-1}$. Notice also that K_0 is the vertex set of K. For $1 \leq t \leq m-1$, let $\mathcal{S}_t = \{\sigma_1, \ldots, \sigma_{r_t}\}$ be the set of t-dimensional faces of K. Observe that

$$K_t = K_{t-1} \cup \mathcal{S}_t.$$

Theorem 12.7 then gives an approach to analyzing the homotopy type of

$$(\underline{C\Omega\Sigma X}, \underline{\Omega\Sigma X})^K$$

by "filtering" it via the spaces $\{(\underline{C\Omega\Sigma X}, \underline{\Omega\Sigma X})^{K_t}\}_{t=0}^{m-1}$.

EXAMPLE 12.10. Another curious example is to start with a simplicial complex K and attach all of its missing faces simultaneously. That is, if $\mathcal{S}$ is the set of all missing faces of K, then let $\overline{K} = K \cup \mathcal{S}$.

Properties when $(\underline{C\Omega\Sigma X}, \underline{\Omega\Sigma X})^K \longrightarrow (\underline{C\Omega\Sigma X}, \underline{\Omega\Sigma X})^{\overline{K}}$ is null homotopic. This is related to Example 12.10. Let K be a simplicial complex on the vertex set $[m]$. For a subset $I \subseteq [m]$ the *full subcomplex* K_I of K is the simplicial complex consisting of those faces in K whose vertices all lie in I. There is a simplicial inclusion $K_I \longrightarrow K$ but this does not have a left inverse that is a simplicial map. On the other hand, the induced map of polyhedral products $(\underline{X}, \underline{A})^{K_I} \longrightarrow (\underline{X}, \underline{A})^K$ does have a left inverse constructed via projection maps [**DS**]. A missing face τ of K has the property that $\partial\tau \subseteq K$ but $\tau \notin K$. If $\tau = (i_1, \ldots, i_k)$, let $I = \{i_1, \ldots, i_k\}$. Then $K_I = \partial\tau$, so $(\underline{X}, \underline{A})^{\partial\tau}$ retracts off $(\underline{X}, \underline{A})^K$. In the case of pairs $(\underline{CX}, \underline{X})$, the polyhedral product $(\underline{CX}, \underline{X})^{\partial\tau}$ is homotopy equivalent to $\Sigma^{k-1} X_{i_1} * \cdots * X_{i_k}$ [**GT2**]. In particular, $(\underline{CX}, \underline{X})^{\partial\tau}$ is not contractible if each of X_{i_1} through X_{i_k} is not contractible. Therefore if $K \longrightarrow L$ is any simplicial inclusion and K and L share a missing face τ then $(\underline{CX}, \underline{X})^{\partial\tau}$ is a nontrivial retract of both $(\underline{CX}, \underline{X})^K$ and $(\underline{CX}, \underline{X})^L$. Hence if the induced map of polyhedral products $(\underline{CX}, \underline{X})^K \longrightarrow (\underline{CX}, \underline{X})^L$ is null homotopic then it must be the case that every missing face of K is a face of L. Consequently, there must be a factorization $K \longrightarrow \overline{K} \longrightarrow L$ where $\overline{K} = K \cup \mathcal{S}$ for $\mathcal{S}$ the set of all missing faces of K.

This lets us focus on when the map $(\underline{CX},\underline{X})^K \longrightarrow (\underline{CX},\underline{X})^{\overline{K}}$ is null homotopic. Note that while it is necessary to fill in the missing faces of K to obtain such a null homotopy of polyhedral products it may not be sufficient. An example when it is sufficient is the following. Take $m = 3$ and let $K = \{\{1\},\{2\},\{3\}\}$ be the simplicial complex determined by the three vertices. The missing faces of K are $\mathcal{S} = \{(1,2),(1,3),(2,3)\}$. Let $\overline{K} = K \cup \mathcal{S}$, so $\overline{K}$ is the boundary of Δ^2. Then, as in [**GT1**], $(\underline{CX},\underline{X})^K \longrightarrow (\underline{CX},\underline{X})^{\overline{K}}$ is null homotopic.

Specialize now to the case when $(\underline{CX},\underline{X})$ is of the form $(\underline{C\Omega\Sigma X},\underline{\Omega\Sigma X})$. In Proposition 12.11 it is shown that the homotopy types of $\Sigma(\underline{C\Omega\Sigma X},\underline{\Omega\Sigma X})^K$ and $(\underline{C\Omega\Sigma X},\underline{\Omega\Sigma X})^{\overline{K}}$ are, in a precise sense, complementary.

PROPOSITION 12.11. *Let K be a simplicial complex on the vertex set $[m]$ and let $\mathcal{S}$ be the set of missing faces of K. Suppose that $\overline{K} = K \cup \mathcal{S}$ has the property that the map of polyhedral products $(\underline{C\Omega\Sigma X},\underline{\Omega\Sigma X})^K \longrightarrow (\underline{C\Omega\Sigma X},\underline{\Omega\Sigma X})^{\overline{K}}$ is null homotopic. Then there is a homotopy equivalence*

$$\Sigma\left(\bigvee_{k=0}^{\infty} \bigvee_{1\leq i_1\leq\cdots\leq i_k\leq m} (X_{i_1}\wedge\cdots\wedge X_{i_k})\wedge\Sigma A\right) \simeq (\underline{C\Omega\Sigma X},\underline{\Omega\Sigma X})^{\overline{K}} \vee \Sigma(\underline{C\Omega\Sigma X},\underline{\Omega\Sigma X})^K.$$

PROOF. By Theorem 12.7 there is a homotopy cofibration

$$\bigvee_{k=0}^{\infty} \bigvee_{1\leq i_1\leq\cdots\leq i_k\leq m} (X_{i_1}\wedge\cdots\wedge X_{i_k})\wedge\Sigma A \xrightarrow{\zeta} (\underline{C\Omega\Sigma X},\underline{\Omega\Sigma X})^K \longrightarrow (\underline{C\Omega\Sigma X},\underline{\Omega\Sigma X})^{\overline{K}}.$$

By hypothesis, the right map is null homotopic. The assertion now follows immediately. □

One condition implying that the map $(\underline{C\Omega\Sigma X},\underline{\Omega\Sigma X})^K \longrightarrow (\underline{C\Omega\Sigma X},\underline{\Omega\Sigma X})^{\overline{K}}$ is null homotopic is if the map ζ in the homotopy cofibration of Theorem 12.7 has a right homotopy inverse. In that case the congruence in Theorem 12.7 allows for more to be said.

PROPOSITION 12.12. *Let K be a simplicial complex on the vertex set $[m]$, let $\mathcal{S}$ be the set of missing faces of K and let $\overline{K} = K \cup \mathcal{S}$. If the map ζ in Theorem* 12.7 *has a right homotopy inverse then the map $(\underline{C\Omega\Sigma X},\underline{\Omega\Sigma X})^K \xrightarrow{p} (\underline{\Sigma X},\underline{*})^K$ factors through the wedge sum of Whitehead products*

$$\bigvee_{k=1}^{\infty} \bigvee_{1\leq i_1\leq\cdots\leq i_k\leq m} [v_{i_1},[v_{i_2},[\cdots[v_{i_k},f]]$$

and Proposition 12.11 *holds.*

PROOF. Let

$$s\colon (\underline{C\Omega\Sigma X},\underline{\Omega\Sigma X})^K \longrightarrow \bigvee_{k=0}^{\infty} \bigvee_{1\leq i_1\leq\cdots\leq i_k\leq m} (X_{i_1}\wedge\cdots\wedge X_{i_k})\wedge\Sigma A$$

be a right homotopy inverse of ζ. Consider the composite

$$(\underline{C\Omega\Sigma X},\underline{\Omega\Sigma X})^K \xrightarrow{s} \bigvee_{k=0}^{\infty} \bigvee_{1\leq i_1\leq\cdots\leq i_k\leq m} (X_{i_1}\wedge\cdots\wedge X_{i_k})\wedge\Sigma A \xrightarrow{\zeta'} (\underline{C\Omega\Sigma X},\underline{\Omega\Sigma X})^K.$$

Since ζ and ζ' are congruent, they have the same image in homology. Therefore, $(\zeta' \circ s)_* = (\zeta \circ s)_*$. As s is a right homotopy inverse of ζ, the map $(\zeta \circ s)_*$ is the identity map in homology. Thus $\zeta' \circ s$ induces an isomorphism in homology and so is a homotopy equivalence by Whitehead's Theorem. Consequently, the homotopy commutative diagram involving ζ' in Theorem 12.11 implies that p factors through the sum of Whitehead products $\bigvee_{k=1}^{\infty} \bigvee_{1\le i_1\le\cdots\le i_k\le m}[v_{i_1},[v_{i_2},[\cdots[v_{i_k},f]]$.

Next, by Theorem 12.7 there is a homotopy cofibration

$$\bigvee_{k=0}^{\infty} \bigvee_{1\le i_1\le\cdots\le i_k\le m} (X_{i_1}\wedge\cdots\wedge X_{i_k})\wedge\Sigma A \xrightarrow{\zeta} (\underline{C\Omega\Sigma X}, \underline{\Omega\Sigma X})^K \longrightarrow (\underline{C\Omega\Sigma X}, \underline{\Omega\Sigma X})^{\overline{K}}.$$

The existence of a right homotopy inverse for ζ implies that the right map in this homotopy cofibration is null homotopic. Therefore Proposition 12.11 holds as well. □

Propositions 12.11 and 12.12 raise several interesting questions.

PROBLEM 12.13. For which K and $\overline{K}$ is the map

$$(\underline{C\Omega\Sigma X}, \underline{\Omega\Sigma X})^K \longrightarrow (\underline{C\Omega\Sigma X}, \underline{\Omega\Sigma X})^{\overline{K}}$$

null homotopic?

PROBLEM 12.14. For which K and $\overline{K}$ does the map ζ in Proposition 12.11 have a right homotopy inverse?

PROBLEM 12.15. In the homotopy decomposition in Proposition 12.11, does each of the wedge summands $\Sigma X_{i_1} \wedge \cdots \wedge X_{i_k} \wedge \Sigma A$ map wholly to one of $(\underline{C\Omega\Sigma X}, \underline{\Omega\Sigma X})^{\overline{K}}$ or $\Sigma(\underline{C\Omega\Sigma X}, \underline{\Omega\Sigma X})^K$, or are there cases when there is a non-trivial decomposition

$$\Sigma X_{i_1} \wedge \cdots \wedge X_{i_k} \wedge \Sigma A \simeq B \vee C$$

with B retracting off $(\underline{C\Omega\Sigma X}, \underline{\Omega\Sigma X})^{\overline{K}}$ and C retracting off $\Sigma(\underline{C\Omega\Sigma X}, \underline{\Omega\Sigma X})^K$? For which $(i_1, \ldots, i_k)$ does $\Sigma X_{i_1} \wedge\cdots\wedge X_{i_k} \wedge \Sigma A$ map wholly into $(\underline{C\Omega\Sigma X}, \underline{\Omega\Sigma X})^{\overline{K}}$ or into $\Sigma(\underline{C\Omega\Sigma X}, \underline{\Omega\Sigma X})^K$?

Despite the potential ambiguity involved in the homotopy decomposition in Proposition 12.11 stated in Problem 12.15, there are cases where interesting information can be extracted regardless. Suppose that for $1 \le i \le m$ each space X_i is a sphere. By definition, the space A is a wedge sum of spaces of the form $X_{i_1} \wedge\cdots\wedge X_{i_\ell}$, and so is homotopy equivalent to a wedge of spheres. Therefore each of the spaces $X_{i_1} \wedge\cdots\wedge X_{i_k} \wedge \Sigma A$ is homotopy equivalent to a wedge of spheres, and hence $\bigvee_{k=0}^{\infty} \bigvee_{1\le i_1\le\cdots\le i_k\le m} (X_{i_1} \wedge \cdots \wedge X_{i_k}) \wedge \Sigma A$ is homotopy equivalent to a wedge of spheres. Any retract of this large wedge is then homotopy equivalent to a wedge of spheres. In particular, from the decomposition in Proposition 12.11 we obtain the following.

COROLLARY 12.16. *Let K be a simplicial complex on the vertex set $[m]$ and let $\mathcal{S}$ be the set of missing faces of K. Suppose that $\overline{K} = K \cup \mathcal{S}$ has the property that the map of polyhedral products $(\underline{C\Omega\Sigma X}, \underline{\Omega\Sigma X})^K \longrightarrow (\underline{C\Omega\Sigma X}, \underline{\Omega\Sigma X})^{\overline{K}}$ is null homotopic. If each space X_i is a sphere for $1 \le i \le m$, then $(\underline{C\Omega\Sigma X}, \underline{\Omega\Sigma X})^{\overline{K}}$ is homotopy equivalent to a wedge of spheres.* □

More is true. If the map $(\underline{C\Omega\Sigma X}, \underline{\Omega\Sigma X})^K \longrightarrow (\underline{C\Omega\Sigma X}, \underline{\Omega\Sigma X})^{\overline{K}}$ null homotopic then in the homotopy cofibration

$$\bigvee_{k=0}^{\infty} \bigvee_{1\leq i_1\leq\cdots\leq i_k\leq m} (X_{i_1}\wedge\cdots\wedge X_{i_k})\wedge\Sigma A \xrightarrow{\zeta} (\underline{C\Omega\Sigma X}, \underline{\Omega\Sigma X})^K \longrightarrow (\underline{C\Omega\Sigma X}, \underline{\Omega\Sigma X})^{\overline{K}}$$

the map ζ induces an epimorphism in homology. If each X_i a sphere for $1 \leq i \leq m$ then $\bigvee_{k=0}^{\infty} \bigvee_{1\leq i_1\leq\cdots\leq i_k\leq m} (X_{i_1}\wedge\cdots\wedge X_{i_k})\wedge\Sigma A$ is homotopy equivalent to a wedge of spheres so ζ_* being an epimorphism implies that a subwedge W may be chosen so the composite

$$W \hookrightarrow \bigvee_{k=0}^{\infty} \bigvee_{1\leq i_1\leq\cdots\leq i_k\leq m} (X_{i_1}\wedge\cdots\wedge X_{i_k})\wedge\Sigma A \xrightarrow{\zeta} (\underline{C\Omega\Sigma X}, \underline{\Omega\Sigma X})^K$$

induces an isomorphism in homology and so is a homotopy equivalence. Thus ζ has a right homotopy inverse, and now Proposition 12.12 applies. Moreover, as ζ' is congruent to ζ by Theorem 12.7, they have the same image in homology, so the composite

$$W \hookrightarrow \bigvee_{k=0}^{\infty} \bigvee_{1\leq i_1\leq\cdots\leq i_k\leq m} (X_{i_1}\wedge\cdots\wedge X_{i_k})\wedge\Sigma A \xrightarrow{\zeta'} (\underline{C\Omega\Sigma X}, \underline{\Omega\Sigma X})^K$$

is a homotopy equivalence and the statement on "factoring through" a wedge sum of Whitehead products in Proposition 12.12 becomes "is" a wedge sum of Whitehead products.

COROLLARY 12.17. *Let K be a simplicial complex on the vertex set $[m]$ and let $\mathcal{S}$ be the set of missing faces of K. Suppose that $\overline{K} = K \cup \mathcal{S}$ has the property that the map of polyhedral products $(\underline{C\Omega\Sigma X}, \underline{\Omega\Sigma X})^K \longrightarrow (\underline{C\Omega\Sigma X}, \underline{\Omega\Sigma X})^{\overline{K}}$ is null homotopic. If X_i is a sphere for $1 \leq i \leq m$ then $(\underline{C\Omega\Sigma X}, \underline{\Omega\Sigma X})^K$ and $(\underline{C\Omega\Sigma X}, \underline{\Omega\Sigma X})^{\overline{K}}$ are both homotopy equivalent to wedges of spheres and the map $(\underline{C\Omega\Sigma X}, \underline{\Omega\Sigma X})^K \xrightarrow{p} (\underline{\Sigma X}, \underline{*})^K$ is a subwedge of the wedge sum of Whitehead products $\bigvee_{k=1}^{\infty} \bigvee_{1\leq i_1\leq\cdots\leq i_k\leq m}[v_{i_1}, [v_{i_2}, [\cdots [v_{i_k}, f]]$.* □

Carrying on, the retraction of S^1 off ΩS^2 induces a retraction of the pair (CS^1, S^1) off the pair $(C\Omega S^2, \Omega S^2)$. Hence for any simplicial complex K we obtain a retraction of $(\underline{CS^1}, \underline{S^1})^K$ off $(\underline{C\Omega S^2}, \underline{\Omega S^2})^K$. Further, this retraction is natural for maps of simplicial complexes. Writing (CS^1, S^1) in the more standard way as (D^2, S^1), the polyhedral product $(\underline{D^2}, \underline{S^1})^K$ is the *moment-angle complex* that is critical to toric topology, more commonly written as $\mathcal{Z}_K$. In the context of Corollary 12.16, we obtain compatible retractions of $\mathcal{Z}_K$ and $\mathcal{Z}_{\overline{K}}$ off $(\underline{C\Omega S^2}, \underline{\Omega S^2})^K$ and $(\underline{C\Omega S^2}, \underline{\Omega S^2})^{\overline{K}}$ respectively. The compatible retractions implies that as the map $(\underline{C\Omega S^2}, \underline{\Omega S^2})^K \longrightarrow (\underline{C\Omega S^2}, \underline{\Omega S^2})^{\overline{K}}$ is null homotopic, so is the map $\mathcal{Z}_K \longrightarrow \mathcal{Z}_{\overline{K}}$. As $(\underline{C\Omega S^2}, \underline{\Omega S^2})^K$ and $(\underline{C\Omega S^2}, \underline{\Omega S^2})^{\overline{K}}$ are homotopy equivalent to wedges of spheres so are $\mathcal{Z}_K$ and $\mathcal{Z}_{\overline{K}}$.

COROLLARY 12.18. *Let K be a simplicial complex on the vertex set $[m]$ and let $\mathcal{S}$ be the set of missing faces of K. Suppose that $\overline{K} = K \cup \mathcal{S}$ has the property that the map of polyhedral products $(\underline{C\Omega S^2}, \underline{\Omega S^2})^K \longrightarrow (\underline{C\Omega S^2}, \underline{\Omega S^2})^{\overline{K}}$ is null homotopic. Then the map $\mathcal{Z}_K \longrightarrow \mathcal{Z}_{\overline{K}}$ is null homotopic, both $\mathcal{Z}_K$ and $\mathcal{Z}_{\overline{K}}$ are homotopy* *equivalent to wedges of spheres, and the map* $\mathcal{Z}_K \longrightarrow (\underline{\mathbb{C}P^\infty}, \underline{*})^K$ *is a subwedge of the wedge sum of Whitehead products* $\bigvee_{k=1}^{\infty} \bigvee_{1 \leq i_1 \leq \cdots \leq i_k \leq m} [v_{i_1}, [v_{i_2}, [\cdots [v_{i_k}, f]]$. □

Corollary 12.18 is connected to important problems in toric topology and combinatorics. By [**BP**] the space $\mathcal{Z}_K$ is homotopy equivalent to the complement of the complex coordinate subspace determined by K. A major question is combinatorics is to determine for which K these complements of coordinate subspace arrangements are homotopy equivalent to a wedge of spheres. A series of papers [**GT1**, **GT2**, **GW**, **IK1**, **IK2**] identified families of simplicial complexes K for which $\mathcal{Z}_K$ is homotopy equivalent to a wedge of spheres, including shifted complexes and those whose Alexander duals are vertex decomposable, shellable or sequentially Cohen-Macauley. All of these are subsumed by what [**IK2**] calls totally fillable or totally homology fillable complexes. Another family of simplicial complexes for which $\mathcal{Z}_K$ is homotopy equivalent to a wedge of spheres is flag complexes whose 1-skeleton is a chordal graph [**GPTW**]. Several papers have examined when the map from $\mathcal{Z}_K$ to $(\underline{\mathbb{C}P^\infty}, \underline{*})^K$ is described by Whitehead products [**AP**, **GT3**, **IK3**].

We end by posing a problem regarding how large might be the family of simplicial complexes with the property that $\mathcal{Z}_K$ is homotopy equivalent to a wedge of spheres. Let $\mathcal{F}$ be the collection of simplicial complexes that are either totally fillable or flag complexes having a 1-skeleton that is a chordal graph.

PROBLEM 12.19. Are there examples of K and $\overline{K}$ in Corollary 12.18 for which $\mathcal{Z}_{\overline{K}}$ or $\mathcal{Z}_K$ is homotopy equivalent to a wedge of spheres but $\overline{K}$ or K is not in $\mathcal{F}$?

Bibliography

[AP] S. A. Abramyan and T. E. Panov, *Higher Whitehead products for moment-angle complexes and substitutions of simplicial complexes* (Russian, with Russian summary), Tr. Mat. Inst. Steklova **305** (2019), no. Algebraicheskaya Topologiya Kombinatorika i Matematicheskaya Fizika, 7–28, DOI 10.4213/tm3995; English transl., Proc. Steklov Inst. Math. **305** (2019), no. 1, 1–21. MR4017598

[A] D. J. Anick, *Homotopy exponents for spaces of category two*, Algebraic topology (Arcata, CA, 1986), Lecture Notes in Math., vol. 1370, Springer, Berlin, 1989, pp. 24–52, DOI 10.1007/BFb0085216. MR1000365

[BBCG] A. Bahri, M. Bendersky, F. R. Cohen, and S. Gitler, *The polyhedral product functor: a method of decomposition for moment-angle complexes, arrangements and related spaces*, Adv. Math. **225** (2010), no. 3, 1634–1668, DOI 10.1016/j.aim.2010.03.026. MR2673742

[Ba] S. Basu, *The homotopy type of the loops on $(n-1)$-connected $(2n+1)$-manifolds*, Algebraic topology and related topics, Trends Math., Birkhäuser/Springer, Singapore, 2019, pp. 1–25, DOI 10.1007/978-981-13-5742-8_2. MR3991174

[BB] S. Basu and S. Basu, *Homotopy groups of highly connected manifolds*, Adv. Math. **337** (2018), 363–416, DOI 10.1016/j.aim.2018.08.006. MR3853054

[BT1] P. Beben and S. Theriault, *The loop space homotopy type of simply-connected four-manifolds and their generalizations*, Adv. Math. **262** (2014), 213–238, DOI 10.1016/j.aim.2014.05.015. MR3228428

[BT2] P. Beben and S. Theriault, *Homotopy groups of highly connected Poincaré duality complexes*, Doc. Math. **27** (2022), 183–212. MR4398609

[BW] P. Beben and J. Wu, *The homotopy type of a Poincaré duality complex after looping*, Proc. Edinb. Math. Soc. (2) **58** (2015), no. 3, 581–616, DOI 10.1017/S0013091515000048. MR3391363

[BM] F. Bosio and L. Meersseman, *Real quadrics in $\mathbf{C}^n$, complex manifolds and convex polytopes*, Acta Math. **197** (2006), no. 1, 53–127, DOI 10.1007/s11511-006-0008-2. MR2285318

[BP] V. M. Buchstaber and T. E. Panov, *Torus actions and their applications in topology and combinatorics*, University Lecture Series, vol. 24, American Mathematical Society, Providence, RI, 2002, DOI 10.1090/ulect/024. MR1897064

[Bu] P. Bubenik, *Free and semi-inert cell attachments*, Trans. Amer. Math. Soc. **357** (2005), no. 11, 4533–4553, DOI 10.1090/S0002-9947-05-03989-9. MR2156720

[C] F. R. Cohen, *Applications of loop spaces to some problems in topology*, Advances in homotopy theory (Cortona, 1988), London Math. Soc. Lecture Note Ser., vol. 139, Cambridge Univ. Press, Cambridge, 1989, pp. 11–20, DOI 10.1017/CBO9780511662614.004. MR1055864

[CMN] F. R. Cohen, J. C. Moore, and J. A. Neisendorfer, *Exponents in homotopy theory*, Algebraic topology and algebraic K-theory (Princeton, N.J., 1983), Ann. of Math. Stud., vol. 113, Princeton Univ. Press, Princeton, NJ, 1987, pp. 3–34. MR921471

[CLOT] O. Cornea, G. Lupton, J. Oprea, and D. Tanré, *Lusternik-Schnirelmann category*, Mathematical Surveys and Monographs, vol. 103, American Mathematical Society, Providence, RI, 2003, DOI 10.1090/surv/103. MR1990857

[DS] G. Denham and A. I. Suciu, *Moment-angle complexes, monomial ideals and Massey products*, Pure Appl. Math. Q. **3** (2007), no. 1, Special Issue: In honor of Robert D. MacPherson., 25–60, DOI 10.4310/PAMQ.2007.v3.n1.a2. MR2330154

[DL] A. Dold and R. Lashof, *Principal quasi-fibrations and fibre homotopy equivalence of bundles*, Illinois J. Math. **3** (1959), 285–305. MR101521

[FHT] Y. Félix, S. Halperin, and J.-C. Thomas, *Loop space homology of spaces of LS category one and two*, Math. Ann. **287** (1990), no. 3, 377–386, DOI 10.1007/BF01446900. MR1060681

[FT1] Y. Félix and J. C. Thomas, *Effet d'un attachement cellulaire dans l'homologie de l'espace des lacets* (French, with English summary), Ann. Inst. Fourier (Grenoble) **39** (1989), no. 1, 207–224. MR1011984

[FT2] Y. Félix and J.-C. Thomas, *Loop space homology of spaces of small category*, Trans. Amer. Math. Soc. **338** (1993), no. 2, 711–721, DOI 10.2307/2154425. MR1134757

[Ga] T. Ganea, *A generalization of the homology and homotopy suspension*, Comment. Math. Helv. **39** (1965), 295–322, DOI 10.1007/BF02566956. MR179791

[Gr] B. Gray, *Abelian properties of Anick spaces*, Mem. Amer. Math. Soc. **246** (2017), no. 1162, v+111, DOI 10.1090/memo/1162. MR3605383

[GIPS] J. Grbić, G. Simmons, M. Ilyasova, and T. Panov, *One-relator groups and algebras related to polyhedral products*, Proc. Roy. Soc. Edinburgh Sect. A **152** (2022), no. 1, 128–147, DOI 10.1017/prm.2020.101. MR4383244

[GPTW] J. Grbić, T. Panov, S. Theriault, and J. Wu, *The homotopy types of moment-angle complexes for flag complexes*, Trans. Amer. Math. Soc. **368** (2016), no. 9, 6663–6682, DOI 10.1090/tran/6578. MR3461047

[GT1] J. Grbić and S. Theriault, *The homotopy type of the complement of a coordinate subspace arrangement*, Topology **46** (2007), no. 4, 357–396, DOI 10.1016/j.top.2007.02.006. MR2321037

[GT2] J. Grbić and S. Theriault, *The homotopy type of the polyhedral product for shifted complexes*, Adv. Math. **245** (2013), 690–715, DOI 10.1016/j.aim.2013.05.002. MR3084441

[GT3] J. Grbić and S. Theriault, *Homotopy theory in toric topology* (Russian, with Russian summary), Uspekhi Mat. Nauk **71** (2016), no. 2(428), 3–80, DOI 10.4213/rm9704; English transl., Russian Math. Surveys **71** (2016), no. 2, 185–251. MR3507473

[GW] V. Grujić and V. Welker, *Moment-angle complexes of pairs (D^n, S^{n-1}) and simplicial complexes with vertex-decomposable duals*, Monatsh. Math. **176** (2015), no. 2, 255–273, DOI 10.1007/s00605-014-0698-z. MR3302158

[HaL] S. Halperin and J.-M. Lemaire, *Suites inertes dans les algèbres de Lie graduées ("Autopsie d'un meurtre. II")* (French), Math. Scand. **61** (1987), no. 1, 39–67, DOI 10.7146/math.scand.a-12190. MR929396

[HST] Y. Hao, Q. Sun, and S. Theriault, *Moore's conjecture for polyhedral products*, Math. Proc. Cambridge Philos. Soc. **167** (2019), no. 1, 23–33, DOI 10.1017/s0305004118000154. MR3968818

[HeL] K. Hess and J.-M. Lemaire, *Nice and lazy cell attachments*, J. Pure Appl. Algebra **112** (1996), no. 1, 29–39, DOI 10.1016/0022-4049(95)00131-X. MR1402394

[IK1] K. Iriye and D. Kishimoto, *Decompositions of polyhedral products for shifted complexes*, Adv. Math. **245** (2013), 716–736, DOI 10.1016/j.aim.2013.05.003. MR3084442

[IK2] K. Iriye and D. Kishimoto, *Fat-wedge filtration and decomposition of polyhedral products*, Kyoto J. Math. **59** (2019), no. 1, 1–51, DOI 10.1215/21562261-2017-0038. MR3934622

[IK3] K. Iriye and D. Kishimoto, *Whitehead products in moment-angle complexes*, J. Math. Soc. Japan **72** (2020), no. 4, 1239–1257, DOI 10.2969/jmsj/82708270. MR4165931

[J1] I. M. James, *Reduced product spaces*, Ann. of Math. (2) **62** (1955), 170–197, DOI 10.2307/2007107. MR73181

[J2] I. M. James, *The suspension triad of a sphere*, Ann. of Math. (2) **63** (1956), 407–429, DOI 10.2307/1970011. MR79263

[K] J. R. Klein, *Poincaré duality spaces*, Surveys on surgery theory, Vol. 1, Ann. of Math. Stud., vol. 145, Princeton Univ. Press, Princeton, NJ, 2000, pp. 135–165. MR1747534

[M] M. Mather, *Pull-backs in homotopy theory*, Canadian J. Math. **28** (1976), no. 2, 225–263, DOI 10.4153/CJM-1976-029-0. MR402694

[L] J. Long, Thesis, Princeton University, 1978.

[N1] J. Neisendorfer, *Primary homotopy theory*, Mem. Amer. Math. Soc. **25** (1980), no. 232, iv+67, DOI 10.1090/memo/0232. MR567801

[N2] J. A. Neisendorfer, *3-primary exponents*, Math. Proc. Cambridge Philos. Soc. **90** (1981), no. 1, 63–83, DOI 10.1017/S0305004100058539. MR611286

[N3] J. A. Neisendorfer, *The exponent of a Moore space*, Algebraic topology and algebraic K-theory (Princeton, N.J., 1983), Ann. of Math. Stud., vol. 113, Princeton Univ. Press, Princeton, NJ, 1987, pp. 35–71. MR921472

[N4] J. Neisendorfer, *Algebraic methods in unstable homotopy theory*, New Mathematical Monographs, vol. 12, Cambridge University Press, Cambridge, 2010, DOI 10.1017/CBO9780511691638. MR2604913

[NS] J. A. Neisendorfer and P. S. Selick, *Some examples of spaces with or without exponents*, Current trends in algebraic topology, Part 1 (London, Ont., 1981), CMS Conf. Proc., vol. 2, Amer. Math. Soc., Providence, R.I., 1982, pp. 343–357. MR686124

[Se] P. Selick, *On conjectures of Moore and Serre in the case of torsion-free suspensions*, Math. Proc. Cambridge Philos. Soc. **94** (1983), no. 1, 53–60, DOI 10.1017/S0305004100060916. MR704802

[St] R. Stöcker, On the structure of 5-dimensional Poincaré Duality complexes, *Comment. Math. Helv.* **57** (1982), 481-520.

[Th] S. D. Theriault, *Homotopy exponents of mod* 2^r *Moore spaces*, Topology **47** (2008), no. 6, 369–398, DOI 10.1016/j.top.2007.09.002. MR2427732

[To] H. Toda, *On the double suspension* E^2, J. Inst. Polytech. Osaka City Univ. Ser. A **7** (1956), 103–145. MR92968

Editorial Information

To be published in the *Memoirs*, a paper must be correct, new, nontrivial, and significant. Further, it must be well written and of interest to a substantial number of mathematicians. Piecemeal results, such as an inconclusive step toward an unproved major theorem or a minor variation on a known result, are in general not acceptable for publication.

Papers appearing in *Memoirs* are generally at least 80 and not more than 200 published pages in length. Papers less than 80 or more than 200 published pages require the approval of the Managing Editor of the Transactions/Memoirs Editorial Board. Published pages are the same size as those generated in the style files provided for $\mathcal{A}\mathcal{M}\mathcal{S}$-LaTeX.

Information on the backlog for this journal can be found on the AMS website starting from `http://www.ams.org/memo`.

A Consent to Publish is required before we can begin processing your paper. After a paper is accepted for publication, the Providence office will send a Consent to Publish and Copyright Agreement to all authors of the paper. By submitting a paper to the *Memoirs*, authors certify that the results have not been submitted to nor are they under consideration for publication by another journal, conference proceedings, or similar publication.

Information for Authors

Memoirs is an author-prepared publication. Once formatted for print and on-line publication, articles will be published as is with the addition of AMS-prepared frontmatter and backmatter. Articles are not copyedited; however, confirmation copy will be sent to the authors.

Initial submission. The AMS uses Centralized Manuscript Processing for initial submissions. Authors should submit a PDF file using the Initial Manuscript Submission form found at `www.ams.org/submission/memo`, or send one copy of the manuscript to the following address: Centralized Manuscript Processing, MEMOIRS OF THE AMS, 201 Charles Street, Providence, RI 02904-2294 USA. If a paper copy is being forwarded to the AMS, indicate that it is for *Memoirs* and include the name of the corresponding author, contact information such as email address or mailing address, and the name of an appropriate Editor to review the paper (see the list of Editors below).

The paper must contain a *descriptive title* and an *abstract* that summarizes the article in language suitable for workers in the general field (algebra, analysis, etc.). The *descriptive title* should be short, but informative; useless or vague phrases such as "some remarks about" or "concerning" should be avoided. The *abstract* should be at least one complete sentence, and at most 300 words. Included with the footnotes to the paper should be the 2020 *Mathematics Subject Classification* representing the primary and secondary subjects of the article. The classifications are accessible from `www.ams.org/msc/`. The Mathematics Subject Classification footnote may be followed by a list of *key words and phrases* describing the subject matter of the article and taken from it. Journal abbreviations used in bibliographies are listed in the latest *Mathematical Reviews* annual index. The series abbreviations are also accessible from `www.ams.org/msnhtml/serials.pdf`. To help in preparing and verifying references, the AMS offers MR Lookup, a Reference Tool for Linking, at `www.ams.org/mrlookup/`.

Electronically prepared manuscripts. The AMS encourages electronically prepared manuscripts, with a strong preference for $\mathcal{A}\mathcal{M}\mathcal{S}$-LaTeX. To this end, the Society has prepared $\mathcal{A}\mathcal{M}\mathcal{S}$-LaTeX author packages for each AMS publication. Author packages include instructions for preparing electronic manuscripts, samples, and a style file that generates the particular design specifications of that publication series.

Authors may retrieve an author package for *Memoirs of the AMS* from `www.ams.org/journals/memo/memoauthorpac.html`. The *AMS Author Handbook* is available in PDF format from the author package link. The author package can also be obtained free of charge by sending email to `tech-support@ams.org` or from the Publication Division,

American Mathematical Society, 201 Charles St., Providence, RI 02904-2294, USA. When requesting an author package, please specify the publication in which your paper will appear. Please be sure to include your complete mailing address.

After acceptance. The source files for the final version of the electronic manuscript should be sent to the Providence office immediately after the paper has been accepted for publication. The author should also submit a PDF of the final version of the paper to the editor, who will forward a copy to the Providence office.

Accepted electronically prepared files can be submitted via the web at `www.ams.org/submit-book-journal/`, sent via FTP, or sent on CD to the Electronic Prepress Department, American Mathematical Society, 201 Charles Street, Providence, RI 02904-2294 USA. TeX source files and graphic files can be transferred over the Internet by FTP to the Internet node `ftp.ams.org` (130.44.1.100). When sending a manuscript electronically via CD, please be sure to include a message indicating that the paper is for the *Memoirs*.

Electronic graphics. Comprehensive instructions on preparing graphics are available at `www.ams.org/authors/journals.html`. A few of the major requirements are given here.

Submit files for graphics as EPS (Encapsulated PostScript) files. This includes graphics originated via a graphics application as well as scanned photographs or other computer-generated images. If this is not possible, TIFF files are acceptable as long as they can be opened in Adobe Photoshop or Illustrator.

Authors using graphics packages for the creation of electronic art should also avoid the use of any lines thinner than 0.5 points in width. Many graphics packages allow the user to specify a "hairline" for a very thin line. Hairlines often look acceptable when proofed on a typical laser printer. However, when produced on a high-resolution laser imagesetter, hairlines become nearly invisible and will be lost entirely in the final printing process.

Screens should be set to values between 15% and 85%. Screens which fall outside of this range are too light or too dark to print correctly. Variations of screens within a graphic should be no less than 10%.

Any graphics created in color will be rendered in grayscale for the printed version unless color printing is authorized by the Managing Editor and the Publisher. In general, color graphics will appear in color in the online version.

Inquiries. Any inquiries concerning a paper that has been accepted for publication should be sent to `memo-query@ams.org` or directly to the Electronic Prepress Department, American Mathematical Society, 201 Charles St., Providence, RI 02904-2294 USA.

Editors

SELECTED PUBLISHED TITLES IN THIS SERIES

1486 **M. Gekhtman, M. Shapiro, and A. Vainshtein,** A Plethora of Cluster Structures on GL_n, 2024

1485 **David Ayala, Aaron Mazel-Gee, and Nick Rozenblyum,** Stratified Noncommutative Geometry, 2024

1484 **Brandon Doherty, Krzysztof Kapulkin, Zachery Lindsey, and Christian Sattler,** Cubical Models of $(\infty, 1)$-Categories, 2024

1483 **Masao Oi,** Simple Supercuspidal L-Packets of Quasi-Split Classical Groups, 2024

1482 **David Burns and Daniel Macias Castillo,** On Refined Conjectures of Birch and Swinnerton-Dyer Type for Hasse–Weil–Artin L-Series, 2024

1481 **Sonja Cox, Martin Hutzenthaler, and Arnulf Jentzen,** Local Lipschitz Continuity in the Initial Value and Strong Completeness for Nonlinear Stochastic Differential Equations, 2024

1480 **Nicolas Burq and Laurent Thomann,** Almost Sure Scattering for the One Dimensional Nonlinear Schrödinger Equation, 2024

1479 **Eva Elduque, Christian Geske, Moisés Herradón Cueto, Laurenţiu G. Maxim, and Botong Wang,** Mixed Hodge Structures on Alexander Modules, 2024

1478 **Qi Chen, Dongyi Wei, and Zhifei Zhang,** Transition Threshold for the 3D Couette Flow in a Finite Channel, 2024

1477 **Vincent Guirardel, Gilbert Levitt, and Rizos Sklinos,** Towers and the First-Order Theories of Hyperbolic Groups, 2024

1476 **Gaëtan Borot, Vincent Bouchard, Nitin K. Chidambaram, Thomas Creutzig, and Dmitry Noshchenko,** Higher Airy Structures, $\mathcal{W}$ Algebras and Topological Recursion, 2024

1475 **Sam Chow and Niclas Technau,** Littlewood and Duffin–Schaeffer-Type Problems in Diophantine Approximation, 2024

1474 **Christian Blohmann and Alan Weinstein,** Hamiltonian Lie Algebroids, 2024

1473 **Dihua Jiang, Zhilin Luo, and Lei Zhang,** Harmonic Analysis and Gamma Functions on Symplectic Groups, 2024

1472 **Thomas Creutzig, Shashank Kanade, and Robert McRae,** Tensor Categories for Vertex Operator Superalgebra Extensions, 2024

1471 **Roberto Feola and Filippo Giuliani,** Quasi-Periodic Traveling Waves on an Infinitely Deep Perfect Fluid Under Gravity, 2024

1470 **Maxwell Stolarski,** Curvature Blow-up in Doubly-warped Product Metrics Evolving by Ricci Flow, 2024

1469 **Steve Boyer, Cameron McA. Gordon, and Xingru Zhang,** Dehn Fillings of Knot Manifolds Containing Essential Twice-Punctured Tori, 2024

1468 **Alice Hedenlund and John Rognes,** A Multiplicative Tate Spectral Sequence for Compact Lie Group Actions, 2024

1467 **Qi Lü and Xu Zhang,** Optimal Feedback for Stochastic Linear Quadratic Control and Backward Stochastic Riccati Equations in Infinite Dimensions, 2024

1466 **Martin W. Liebeck, Gary M. Seitz, and Donna M. Testerman,** Multiplicity-free Representations of Algebraic Groups, 2024

1465 **Tobias Rossmann and Christopher Voll,** Groups, Graphs, and Hypergraphs: Average Sizes of Kernels of Generic Matrices with Support Constraints, 2024

1464 **Jens Marklof and Andreas Strömbergsson,** Kinetic Theory for the Low-Density Lorentz Gas, 2024

1463 **You Qi and Joshua Sussan,** p-DG Cyclotomic nilHecke Algebras II, 2024